-132

MODERN MATERIALS

Advances in Development and Applications

VOLUME 7

CONTRIBUTORS

J. C. Chaston
Donald H. Leeds
L. D. Locker
Thomas M. Maloney
Arthur L. Mottet
E. C. Shuman
William H. Smith
Lyle F. Yerges

MODERN MATERIALS: ADVANCES IN DEVELOPMENT AND APPLICATIONS is published in cooperation with the American Society for Testing and Materials and the Administrative Committee on Research.

MODERN MATERIALS

ADVANCES IN
DEVELOPMENT AND APPLICATIONS

EDITED BY

BRUCE W. GONSER

Battelle Memorial Institute
Columbus, Ohio

ADVISORY BOARD

J. J. HARWOOD · HENRY H. HAUSNER · E. C. JAHN
IVOR JENKINS · HERMAN MARK · J. T. NORTON
ALF SANENGEN

VOLUME 7
1970

ACADEMIC PRESS
NEW YORK AND LONDON

ACADEMIC PRESS, INC.
111 Fifth Avenue, New York, New York 10003

United Kingdom Edition published by
ACADEMIC PRESS, INC. (LONDON) LTD.
Berkeley Square House, London W1X 6BA

LIBRARY OF CONGRESS CATALOG CARD NUMBER: 58-12811

PRINTED IN THE UNITED STATES OF AMERICA

850

CONTENTS

Particleboard
THOMAS M. MALONEY AND ARTHUR L. MOTTET

Acoustical Materials
LYLE F. YERGES

Materials Produced by Electrical Discharges
L. D. LOCKER

Pyrolytic Graphite
WILLIAM H. SMITH AND DONALD H. LEEDS

Materials for Temperature Measurement

J. C. CHASTON

Thermal Insulation Systems

E. C. SHUMAN

LIST OF CONTRIBUTORS

J. C. CHASTON, *Consultant; Formerly Manager, Research Laboratories, Johnson Matthey & Co., Ltd., London, England*

DONALD H. LEEDS, *Super-Temp Company, Santa Fe Springs, California*

L. D. LOCKER, *Bell Telephone Laboratories, Murray Hill, New Jersey*

THOMAS M. MALONEY, *College of Engineering Research Division, Washington State University, Pullman, Washington*

ARTHUR L. MOTTET, *Research and Development, Long-Bell Division, International Paper Company, Longview, Washington*

E. C. SHUMAN, *Professional Engineer, State College, Pennsylvania*

WILLIAM H. SMITH, *Super-Temp Company, Santa Fe Springs, California*

LYLE F. YERGES, *Lyle F. Yerges, Consulting Engineers, Downers Grove, Illinois*

PREFACE

In continuing this serial publication on *Modern Materials* no general change in coverage has been made. The objective has remained that of presenting diverse subjects in the broad field of materials and of discussions by authoritative specialists for the benefit of nonspecialists. Although this is an age of specialization there is still need to know something of what is going on in other fields. Advances are made on too wide a front in the whole area for any materials engineer to be complacent with the knowledge he has gained in one narrow sector.

Certainly one of the most common and useful materials available is wood. In the first chapter of Volume 1 Carl de Zeeuw covered Some New Developments in Wood as a Material. Chapters on plastics and effects of radiation were covered in later volumes, and in Volume 6 the development of Radiation-Processed Wood-Plastic Materials was described by Stein and Dietz. It has seemed logical to follow this with a discussion of the exciting growth of Particleboard. This has been done by Thomas Maloney of Washington State University and Arthur Mottet of the International Paper Corporation in a most capable manner. Conversion of waste by-products into an extremely useful, profitable, and modern material of construction is not only of interest to materials-minded people but is also a lesson in conservation through research.

Closely associated with modern building and machine construction is the need for knowledge of materials as related to their acoustical properties. Few will disagree with the need for noise suppression, or at least control. Here Lyle Yerges, from his long experience as a consulting engineer in the field, has presented an authoritative discussion of the principles involved and their practical application.

Also, along this same line of association with modern needs in building and machine construction, is thermal insulation. For comfort in normal living, as well as in controlling high temperatures and maintaining low temperatures, thermal insulation is obviously an important consideration and every materials engineer should have some information on the principles involved. E. C. Shuman has covered the many considerations very capably. He was formerly with the institute for Building Research at Pennsylvania State University and has been Chairman of Committee E-6 (Methods of Testing Building Construction) of the American Society for Testing and Materials.

For the fourth chapter an unusual and intriguing subject is introduced with Materials Produced by Electrical Discharges. In a way this may deal as much or more with processes and procedures than with materials alone, but the many applications, present and potential, certainly affect materials. This is both an engineering and scholarly presentation by Dr. L. D. Locker. He was on the faculty of the Engineering Materials Laboratory of the University of Maryland before joining the Bell Telephone Laboratories, and through his researches there became greatly interested in thin film coatings and the potentialities of utilizing electric discharges in effecting coatings and chemical reactions.

In Volume 4 an excellent chapter was presented by Earle Shobert II on the general subject of Carbon and Graphite. Because of the interest shown in one phase of this subject it has seemed desirable to devote another chapter to Pyrolytic Graphite. This specific form of graphite formed by chemical vapor deposition has some remarkable properties. Applications have been particularly in the aerospace field although usefulness may extend to a much wider area where high temperatures and special conditions are encountered that cannot be met by metals and most ceramics. Two recognized specialists in this field have combined to give a comprehensive and authoritative presentation. Dr. William H. Smith is President and Don Leeds is Manager of Development Engineering of the Super-Temp Company. Their comments are based on actual manufacturing and use experience.

Temperature is such an important state in research and in nearly all operations dealing with materials that a chapter devoted to the materials involved in its measurement has seemed logical. For this Dr. J. C. Chaston has assembled material from his many years of experience. Before retiring a few years ago Dr. Chaston was manager of the research laboratories of Johnson Matthey and Company in England, one of the world's leading producers of platinum group metal products. Anyone even mildly interested in materials used for temperature measurement will find much of interest in this informative discussion.

Many more chapters dealing with modern materials could be added since the field is nearly inexhaustible. Because of this wide diversity and the consequent trend toward specialization, however, the *Modern Materials* serial publication will undergo a change of format for future volumes. The new format contemplates a series of monographs in which each work schedule will be devoted entirely to a single topic that heretofore might have been treated in a more circumscribed chapter. In this way, we hope to cover each topic in an exhaustive manner, bringing the interested reader fully up to date on the broadest spectrum of developments in any particular field.

BRUCE W. GONSER

November, 1969

Contents of Previous Volumes

MODERN MATERIALS

Advances in Development and Applications

VOLUME 7

PARTICLEBOARD

Thomas M. Maloney and Arthur L. Mottet

College of Engineering Research Division, Washington State University, Pullman, Washington
and Research and Development, Long-Bell Division, International Paper Company,
Longview, Washington

1

I. Introduction

Particleboard is a relatively new forest product, and its manufacture represents the fastest growing segment of the entire wood industry. Production has climbed in the United States from less than one million to over 1½ billion square feet (¾-inch basis) per year during a period of 15 years. Currently, the rate of increase of productive capacity can only be described as spectacular. One plant in particular can produce over 1200 tons of board per day, making it the largest plant for this product in the world.

A. DEFINITION AND DESCRIPTION

Particleboard is a type of composition board made up largely of particles of wood or similar lignocellulosic material bound together with an adhesive. The binder is usually a thermosetting synthetic resin, most frequently of urea-formaldehyde, but sometimes of the phenol-formaldehyde type. In commercial production, the board is invariably consolidated under heat and pressure.

The most notable feature of this material is its grainless character and the controlled variation in surface smoothness and strength properties possible in different forms of the product. Also notable is the large sizes currently available.

Wood particles constitute by far the larger part of particleboard manufactured, although appreciable quantities are produced with flax shives and bagasse. Other materials such as bark, rice hulls, hemp, jute, and crushed nut shells have also been considered for use in the manufacture of particleboard.

The two major types of particleboard are (1) board produced by forming resin-treated wood particles into a mat and then pressing in a platen hot press; and (2) board produced by extrusion from a die. These differ radically in physical properties, owing principally to the effects of the method of pressing on the orientation of particles in the board.

The platen-pressed boards are characterized generally by much greater dimensional stability in the plane of the board, and the extruded boards by greater tensile strength perpendicular to the face of the board and by somewhat greater dimensional stability in thickness. The growth of manufacturing of mat-formed platen-pressed board has far outstripped that of extruded board, so that only a small fraction of the total particleboard produced at present is extruded. Most of the extruded board is produced in captive plants by furniture manufacturers.

Although this discussion will not include molded particleboard, it should be mentioned that some is being produced in the United States and that this is an important segment of the European industry. An expansion of this branch of the industry is quite possible.

Because the size and shape of wood particles can be varied within a wide range, and the wood species, kind and amount of resin, kind and amount of other additives, pressing conditions, and many other variables are controllable, the character of particleboard can be tailored within a wide range to many differing use requirements. The effects of manufacturing variables are described in Section IV.

B. History

Old patents, predating the American Civil War, testify that the concept of producing a composition or synthetic board by bonding together wood particles is more than a hundred years old. Nevertheless, the particleboard industry is the youngest of the major forest products industries. Economic conditions and the technology of adhesives did not become ripe for a particleboard industry until during and immediately subsequent to World War II. It appears that the first industrial produc-

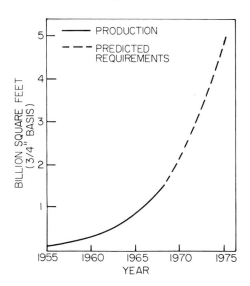

Fig. 1. The growth in production of particleboard and future requirements.

tion of particleboard took place in Germany and Switzerland during the war years, although there is a reference to a Czechoslovakian plant predating 1940. The first American production was initiated in New England in 1945. During the first decade of the existence of the American particleboard industry, growth of production was slow and uneven. The industry advanced rapidly during the second decade of its existence until today (1969), it rates as a major forest products industry and rapid growth continues.

C. PRESENT GROWTH OF THE INDUSTRY

The rapid growth of the industry, presently at about 20% per year, is predicted to continue for some time. The production from 1955 through 1968 and the predicted requirements for particleboard through 1975 are shown in Fig. 1. A threefold increase in requirements to over 5 billion square feet is predicted by 1975, and further predictions forecast the possibility of a 50 billion square feet requirement in the year 2000.

D. RELATION TO OTHER FOREST PRODUCTS INDUSTRIES

The sister industries of particleboard manufacturing are lumber, veneer and plywood, low-density fiberboard (insulating board), medium density fiberboard, and high density fiberboard (hardboard). These, with the paper and paperboard industry, make up the major forest products industries. In the raw material supply picture, the relation of particleboard manufacturing to its sister industries could be called symbiotic. Manufacture of lumber and plywood requires relatively large and expensive logs. The fiberboards and paper and paperboard products generally require pulp chips, which may be produced from lower-cost pulp bolts or roundwood or from solid-wood residue developed in lumber and plywood manufacturing.

Although particleboard can also be made from these raw materials, it is the least restrictive of the wood products mentioned above in its raw material requirements—both in the form and condition of the material as well as in wood species. A large percentage of particleboard production, for example, is based on shavings from lumber planing mills and also on other comminuted forms of wood residue from similar woodworking operations. Another example is aspen roundwood which is very fast growing. Not considered desirable for other uses, it is well suited to particleboard production. As a consequence, particle board raw material cost is generally the lowest in the wood products industries, and particleboard end-product cost in terms of either weight or volume tends also to be the lowest.

II. Manufacturing Methods

The manufacture of particleboard will be described in a generalized perspective first, followed later in Section IV by a brief discussion of process variations that lead to important product variations.

The basic steps in the process are gathering of raw materials, particle preparation, classifying, drying, additive application, mat formation, prepressing, pressing, cooling and conditioning, trimming, and sanding. The layout of a typical plant is shown in Fig. 2.

A. Raw Materials

One factor that has contributed much to the meteoric rise of the particleboard industry is its versatility in the acceptance of a wide variety of raw materials. As mentioned earlier, these range from wood residues developed during manufacture of lumber, plywood, and furniture to agricultural residues such as flax shives and bagasse. However, in the United States, wood residues are plentiful and these form the backbone of the industry. Planer shavings produced during the surfacing of lumber have proved to be the most popular form of raw material because of their availability in large volume at low cost, and because of the ease with which they convert to usable particles.

Solid wood residues such as slabs, edgings, and trim are also satisfactory raw material for particleboard manufacture, although more energy is required to produce a usable particle. Small roundwood such as is developed during forest thinning operations may also be used for this purpose, although the practice is more common in European plants. Sawdust has not been used to any large extent, but increasing usage is

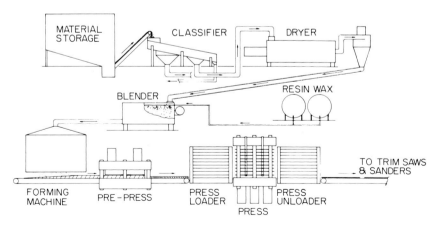

Fig. 2. Schematic layout of a typical particleboard plant.

expected as the supply of shavings diminishes and technological advances are made. Bark, another plentiful residue material in the manufacture of lumber, plywood, and paper, also has found little use since it has low strength properties; but again, increased usage may be foreseen as the disposal problem for this material becomes more acute and as new techniques for its use in particleboard are found. Moreover, demand for more building materials may accelerate the research necessary for utilization of this very inexpensive and available raw material.

A source of raw material previously untapped is the narrow edge trim that is removed during one of the final steps in the manufacture of plywood. Great amounts of this material are available in the plywood-producing areas of the West and South, and it is becoming a basic raw material for some plants. A simple grinding operation reduces such material to a suitable particle.

B. Particle Preparation

The particles are prepared by means of several different machines, depending upon the form of the raw material and the desired geometry of the particle. Particle geometry has a profound influence on board properties and must be carefully controlled in order to assure uniformity of final product. The term "furnish" is used to define the particles that have been prepared for processing into particleboard.

The term "particle" is used in the generic sense to refer to all types of elements involved in board manufacture. The commercial standard (U. S. Department of Commerce) on particleboard, CS236-66, *Mat-Formed Wood Particleboard,* specifies the wood particle requirement as follows: "The wood particles shall be flakes, chips, shavings, slivers, and similar forms that are produced from any natural wood by cutting, hammermilling, grinding and similar processes."

Definitions for various types of particles used in particleboard have been developed by the American Society for Testing and Materials in a standard entitled "Standard Definitions of Terms Relating to Wood-Base Fiber and Particle Panel Materials," ASTM D1554-65. The particles used in particle panel materials are defined as follows:

Chips. Small pieces of wood chopped off a block by ax-like cuts as in a chipper of the paper industry, or produced by mechanical hogs, hammermills, etc.

Curls. Long, flat flakes manufactured by the cutting action of a knife in such a way that they tend to be in the form of a helix.

Fibers. The slender, threadlike elements or groups of wood fibers or similar cellulosic material resulting from chemical or mechanical fiberization, or both, and sometimes referred to as fiber bundles.

Flake. A small wood particle of predetermined dimensions specifically produced as a primary function of specialized equipment of various types, with the cutting action across the grain (either radially, tangentially, or at an angle between), the action being such as to produce a particle of uniform thickness, essentially flat, and having the fiber direction essentially in the plane of the flakes, in overall character resembling a small piece of veneer.

Shaving. A small wood particle of indefinite dimensions developed incidental to certain woodworking operations involving rotary cutterheads usually turning in the direction of the grain; and, because of this cutting action, producing a thin chip of varying thickness, usually feathered along at least one edge and thick at another and usually curled.

Slivers. Particles of nearly square or rectangular cross-section with a length parallel to the grain of the wood of at least four times the thickness.

Strand. A relatively long (with respect to thickness and width) shaving consisting of flat, long bundles of fibers having parallel surfaces.

Wood Wool (Excelsior). Long, curly, slender strands of wool used as an aggregate component for some particleboards.

(Another term and type of particle which has gained some currency in particleboard manufacturing but which has not been recognized in the standards is "wafer." This could be considered to fall under the "flake" classification as a very thick flake.)

The predominant particles used in the industry are shavings, flakes, and fibers. These three basic elements and boards produced from them are illustrated in Fig. 3.

It is important to point out at this juncture a semantic problem of

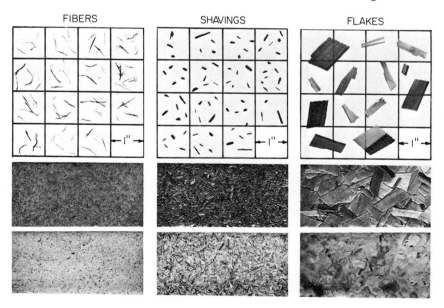

FIBERS   SHAVINGS   FLAKES

FIG. 3. Typical fibers, shavings and disk-cut flakes, and boards made from these wood elements.

nomenclature. Although the term "particle" is used generically, it is also the most descriptive term to apply to an element having no outstanding geometric feature. So it is with the elements shown in the middle row of Fig. 3, which are typical *particles*. If elongated along the grain they are referred to as *slivers*, although not necessarily sharp, and if further elongated they are referred to as *strands*. The end point of this elongation would be excelsior. When width instead of length becomes the geometric feature, the particle assumes the more descriptive term of *flake*. When particles are further reduced in size, the outcome may be either a *fiber*, or simply a smaller particle, usually termed a *fine*. A *fiber* as a particle is actually a group of fibers in the botanical context and is normally produced by grinding of *chips*, the largest of the elements involved in particleboard manufacture. The confusion regarding particles is somewhat alleviated by the fact that these elements are generally produced from shavings, the most preferred raw material. Shavings definitely appear to be particles; hence, although a board comprised of this element could be referred to as a shavings board, the name "particleboard" fits nicely and does not confuse the consumer.

The specific type of particle contributes special properties preferred for particular uses and is thus an important means of characterizing particleboards. Properties of boards will be covered later in Section III.

A simple milling action is sufficient to break oversize shavings into small particles of desirable geometry. Shavings are produced by the knife action in lumber planers traveling parallel to the grain. This action produces thin, featherlike particles that flow easily and uniformly from step to step, and fit well together in the internal structure of the board.

In use of solid wood residues, there are a number of possibilities for preparation of particles suitable for board manufacture. The first and perhaps simplest procedure, though little used, is to reduce the wood first by means of a conventional chipper such as is used in the initial step for pulp and paper manufacture. This chip, from ½ to ¾ inches long, is then subjected to hammermill action to produce small, sticklike particles approximating the shape developed from shavings. An alternative procedure is to pass the chips through a type of ring-knife machine (e.g., Pallmann), which subjects the chips to further knife action such as to produce a particle somewhat similar in appearance to that developed from shavings. However, it is more closely related to the flake, the type of particle next to be discussed.

Flakes can be described as thin, veneerlike elements of controlled thickness and length. These are produced by knife action on solid wood, and they normally range from ¼ to 2 inches in length and from 5 to 30 mils in thickness. Flakes, as produced by disk-type flakers, are usually too wide for efficient handling through subsequent processing steps and for proper intermeshing with other flakes during mat formation; thus, they are sometimes subjected to further milling. The large, thick flakes termed "wafers" are produced on drum-type flakers but are little used for particleboard.

Fibers, the last element to be discussed, usually are produced by grinding chips in an attrition-type mill after first steaming for a period of time at moderate pressures. The steaming softens the lignin in the wood, which is the natural bonding agent of the wood fibers. Fibers more appropriate for most board production are ground in this way, although a limited amount of grinding is done on green or wet chips. The fibers thus produced are actually small groups of the basic fiber of which the wood is composed, and are technically referred to as "fiber bundles." A new machine for grinding material into fiber has been developed wherein the grinding is done under pressure, resulting in a fiber suitable for a wide range of fiberboards. This machine, or refiner, has produced a technological breakthrough in being able to produce suitable fiber from sawdust and opening the door to wider use of mixed species in producing fiber for board manufacture.

Although basic particle types themselves contribute significantly to to board properties, the size or distribution of sizes of particles within any one type also has a profound effect on board properties. This particle size variation, as will be seen presently, is amenable to control. Placement of particles within the board structure can also be manipulated for still more variations in board properties.

C. Drying

After particle preparation, the next step is normally a drying operation that removes all excess water from the particles. This operation is performed in highly efficient dryers that evaporate water from the particles during a very short retention time. Some dryers, for example, are direct fired; that is, the combustion of the fuel takes place within the chamber but adjacent to the particles moving through the dryer. The hot air accomplishing the drying can exceed 1000°F, and the retention time of the particles may be a matter of several seconds. Under these conditions, the wood particles do not burn because of the low retention time and because each particle is surrounded in an envelope of its own moisture. The final moisture content of the particles is normally in the order of 2 to 4%.

D. Classification

The purpose of classifying is to remove the oversize particles, which are returned to the grinding operation, and to segregate the remaining particles into any size classification desired. The very fine dust is sometimes removed and discarded; but, in some plants, this material is allowed to remain in the board. Depending upon the process, classifying may be performed before or after drying. Classifying is accomplished by screening, air separation, or a combination of both.

E. BLENDING OF ADDITIVES

A binder, usually of a thermosetting resin type, is then added to the particles. Either urea-formaldehyde or phenol-formaldehyde types are used, as mentioned earlier, but predominantly the former because of lower cost and faster cure times. Phenol types are used when the product is intended for outdoor use or for secondary manufacture calling for high temperatures such as those found in some laminating processes. Resins are usually used in liquid form, although in some rare instances dry powdered resins are employed. In the liquid application, the resin is applied by spraying into a constantly churning or cascading mass of particles in either batch-type or continuous machines, although virtually all processes are now continuous. The purpose of the binder is to bond all particles together into a solid, cohesive board. It is highly essential, therefore, that some binder be present on all particles in proper amounts to achieve optimum bonding action at minimum cost.

Wax is applied in similar manner either in the form of an emulsion or in the molten state. The purpose of the wax is to provide a measure of water resistance to the board after it is in service. Wax, however, does not impart any resistance to dimensional changes in the board caused by changes in atmospheric humidity. Normally, the resin content ranges from 3 to 8% and the wax from ½ to 1½%. During this operation, the water content of the particles normally increases to about 10 or 12%, the maximum allowable in most processes. Fast press times require lower moisture contents. Occasionally, higher moisture contents are permitted in the surface layers in order to assist in plasticizing the surface particles during pressing and thus achieve a smoother, more stable surface.

Particles may also be treated with other additives such as fire retardants and fungicides for particular purposes. An important consideration in regard to special additives is to insure that they do not adversely affect the bonding action of the resin.

F. MAT FORMATION

After application of binders and other additives, the particles are ready to be formed into a loose mat. This is accomplished by ingenious forming machines, which distribute the particles uniformly, usually onto a metal caul plate. Most formers use some method of dropping the particles upon the caul plate from a height of several feet by means of

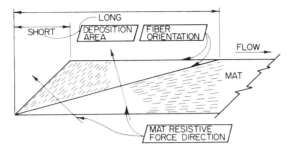

Fig. 4. The particle orientations occurring with long and short deposition angles in formation of mats.

roll metering or air sifting. The amount of material deposited on the caul plate at this point is related to the final thickness desired in the board and its final density. Particles can be layered during mat formation according to particle size by several means. The two most popular types of formers prepare three-layered and multi-layered mats.

In forming the mat, it is important to distribute the particles over a wide area in order to have a "long deposition angle" as shown in Fig. 4. Otherwise, with the "short deposition angle," also illustrated, particles are oriented to the detriment of broad properties. In particular, bending strength can be decreased because the particles are deposited at an angle and the particle length is not used to full advantage. In addition, mats formed with the short deposition angle create a horizontal component to the pressing force of the hot press, which can shift the press crown off its longitudinal center line.

G. Prepressing

Since the mats at this stage are often too thick to fit into the openings between the plates of the press, the mats are usually directed to a prepressing operation. This may be a roll-type press, but is usually a platen press that can reduce the thickness about 50%. Prepressing reduces the space necessary between the press platens and develops sufficient cohesiveness in the mat to permit its transfer to the next step without disrupting the mat. This is extremely important in high-speed, high-production plants where mats must be quickly prepared for pressing. This cohesiveness results from the tackiness of the resin and is an important feature of binders used for this purpose.

The new "caulless" system removes the metal caul plate from the mat-forming process. Mat forming is accomplished on a plastic sheet, and it is always necessary to prepress. This prepressing exerts a pressure of about 400 psi on the mat. Afterwards the mat is removed from the plastic sheet and is then dependent upon its own cohesiveness to retain its integrity.

H. Hot Pressing

The prepressed mats are then accumulated into the press loader, which receives as many mats as the press can accommodate at one time. The speed of the mat forming and prepressing operations must be such that the proper number of mats are accumulated in the press loader during the interval required for pressing the previous load. All mats move simultaneously into the press as the pressed boards are withdrawn from the opposite side of the press. In the older or conventional system, the metal caul carrying the mat also goes into the press. For the newer caul-less systems, the highly prepressed mats are deposited directly onto the press plates. The press is closed quickly in accordance with a prearranged schedule in order to avoid precuring the resin before consolidation occurs. After completion of the pressing cycle, which may range from 2 to 12 minutes, depending upon the thickness of the board and the type of resin used, the press opens, boards are discharged, and the new load of mats is inserted. Most presses are steam heated and plate temperatures are about 325°F. Hot water and hot oil are other heating media used.

Most presses close to "stops," which are metal bars between the press plates corresponding to the board thickness being pressed. Clever new devices are now available that allow pressing to thickness without stops. Elimination of stops is desirable as they tend to become deformed after a period of time, and accumulations of debris build up on them and change the board thickness or caliber.

Multiopening presses may range in size from 4 by 8 feet to 8½ by 33 feet, and in number of openings from 1 to 30. Some single-opening presses may be up to 12 ft wide and 49 ft long. Most plants in operation employ the multiopening presses. The multi-opening systems have found the most favor because of their higher production rates, the single-opening presses being used in special cases or for special products. The essential processing steps described, however, are similar for both multi-opening and single-opening press systems.

I. Conditioning

After hot pressing, boards are either allowed to cool, or are hot stacked to promote further curing of the adhesive before machining to size. The conditioning treatment may also include increasing the moisture content of the board to a level corresponding to ultimate use conditions in order to reduce dimensional change and warp tendencies.

In general, with urea-bonded boards, it is very important to cool briefly after pressing. This can be accomplished in a giant "cooling wheel," where ambient air is passed across both surfaces of the board. If these boards were hot stacked, the resin in the boards could overcure and degrade the board.

J. Sizing of the Finished Boards

Huge belt sanders are used to smooth and dimension the boards to final thickness. Properly manufactured particleboards are virtually warp free because internal stresses are "balanced" from surface to surface. Improper sanding, such as removing a substantially greater amount of material from one surface, could destroy this stress balance and yield warped panels.

Special panel saws cut the pressed boards into the final sizes. Particleboard, because of its high density and the abrasive synthetic resin used for bonding, must be cut with carbide-tipped saws. Conventional steel saws dull rapidly, burn the board edges, and tear out the particles on the corners. Initially, most of the product was cut to the conventional plywood panel size of 4 by 8 feet. However, changes in the uses of particleboard now require a multiplicity of sizes to be available. This has resulted in elaborate "cut-up" plants and intricate systems of loading box cars for maximum space utilization and product protection.

Other refinements in some plants include systems to fill and prime board surfaces, to edge bond particleboard with wood strips or to fill the relatively porous edges with plastic formulations. A recent development has been the treating of the surfaces of floor underlayment with a hot melt coating to prevent the pickup of moisture.

III. Physical Properties

Particleboard can be considered one of a family of wood-based panel materials. The members of this family, besides particle board, are

Lumber, particularly when edge-glued into wide panels
Plywood
Insulating board (or low density fiberboard)
Medium density fiberboard
Hardboard (or hard-pressed fiberboard)

Each of the members of this family has a constellation of physical properties and functional characteristics (or rather a range of properties and characteristics) in combination with a general range of cost levels. Each one is especially suitable for certain uses and less suitable or not suitable at all for other uses.

The most generally recognized procedures for the determination of the physical properties of particleboard in the U. S. and Canada are those specified by the American Association for Testing and Materials in a set of standards designated as ASTM D1037-64.

Needless to say, sound materials engineering requires that particleboard not be used where the other types of panels are more suitable and vice versa. Table I shows how particleboard compares in general with its sister materials.

A. DENSITY

If one property can be taken as most fundamental to the character of particleboard, it is board density. Density per se might be thought to be only important to the weight of the board, but it is also in fact so intimately related to all the other physical properties that density is the first bit of data essential to identifying the general nature of any particular board product.

Although it is possible to produce board outside this range, most commercial particleboard falls within the range of 35 to 55 lb/cu ft. Boards at 37 lb/cu ft or less are rated as low density boards and those at 50 lb or over are rated as high density boards. Everything in between is rated as medium density. Most of the particleboard used is in the medium density category and indeed most of this falls within the range of 40 to 47 lb/cu ft.

All mechanical properties values climb rapidly with increasing board density, everything else being kept constant. To a lesser degree, moisture resistance characteristics are also associated with density. In general, where increased weight is of no concern, the most economical means of attaining a specified level of properties is by increasing density. Where product weight is a consideration, the maximum board density consistent with the requirement of the product will in general permit the most economical utilization of materials.

B. HARDNESS

Of all board properties, hardness, or dent resistance, is most closely tied to board density and most independent of all other factors. It is so closely associated with board density that particleboard users seldom specify a hardness value, this being simply accounted for in the board density specification. This property is measured by determining the load required to embed a steel ball of a specified diameter to a depth of one-half its diameter.

Because particleboards are usually denser than natural wood and less dense than hardboards, the dent resistance of particleboards is generally higher than that of plywood and lumber (on side grain as

TABLE I
COMPARISON OF PARTICLEBOARD WITH OTHER WOOD-BASED PRODUCTS

Properties	Particle-board	Hard-board	Insulating board	Plywood	Edge-glued lumber
Weight	Moderate to heavy	Heavy	Very light	Light to moderate	Light to moderate
Bending strength	Moderate to high	High	Very low	High	High parallel to grain, low across grain
Bending stiffness	Moderate to high	High	Very low	Very high	High parallel to grain, low across grain
Hardness	Moderate to high	High	Low	Moderate	Moderate
Tensile perpendicular ("Internal Bond")	Moderate to high	Low to moderate	Very low	High	High
Thickness dimensional stability	Moderate	Moderate	High	Moderate	Moderate
Linear dimensional stability	Moderate to high	Moderate	Low	High	Very high parallel to grain, low across grain
Cost	Low	Moderate	Very low	High	Very high

distinguished from end grain) and generally lower than that of hard-board. The hardness of particleboards, determined by the Janka Ball Method, is approximately as follows:

High Density	1300 to 2000 lb
Medium Density	750 to 1500
Low Density	400 to 800

C. Modulus of Rupture

Bending strength of board is the most frequently specified mechanical property of particleboard. Although the breaking strength of particleboard in bending is very seldom of direct concern in utilization of particleboard, this property is quickly and easily determined and is connected in a rough way with all the other strength properties. The modulus of rupture values generally range as follows:

High Density	3000 to 8000 psi
Medium Density	1600 to 5000
Low Density	800 to 2000

Like hardness, the bending strength is greatly influenced by board density within any of the density classifications. But, unlike hardness, it is also strongly affected by factors such as wood species, particle geometry, kind and amount of resin binder, and other process factors.

D. Modulus of Elasticity

The modulus of elasticity of particleboard, its stiffness or resistance to deflection when subjected to bending stresses, is of considerable importance in many uses of particleboard. The greater the stiffness and the thinner the board required to provide a required degree of rigidity, the more attractive will be the performance/cost ratio of a board. Examples of uses in which stiffness can be of paramount importance are shelving, table tops, many case goods parts, wall paneling (particularly when applied directly to studs), flooring underlayment, and mobile home floors. Modulus of elasticity levels that can be expected in commercial particleboard are as follows:

High Density	350,000 to 1,000,000 psi
Medium Density	250,000 to 500,000
Low Density	150,000 to 300,000

E. Tension Perpendicular to Surface

This property, more frequently designated by the less precise term, "internal bond," is of critical importance in many applications of particleboard. The property is determined by gluing metal or wood blocks to the faces of a 2-by-2-inch specimen of the board and then measuring the load required to pull the blocks apart in a direction perpendicular to the faces of the board. This in effect measures the resistance of the board to delaminating forces. In a high proportion of uses, particleboard is overlaid with plastic laminates or films, or hardwood veneers. The resistance of these overlays to separation from the particleboard substrate is dependent in part on the internal bond strength of the board, particularly in the zone immediately adjacent to the overlay.

Although internal bond is of more direct importance than modulus of rupture in the performance characteristics of a board, and is perhaps an even more sensitive indicator of all the other mechanical properties, it is more laborious and time-consuming to test and therefore is not as frequently tested.

The range of values encountered in commercial particleboard is roughly as follows:

High Density	125 to 500 psi
Medium Density	60 to 400
Low Density	20 to 40

F. Screw-Holding Strength

Because so many of the industrial uses of particleboard involve fastening with screws, for example fastening of hinges on a kitchen cabinet door or a phonograph case, screw-holding specifications are frequently employed. This mechanical property is determined by measuring the force required to extract from the board a screw of specified size and shape (No. 10 sheet-metal screw) inserted under strictly standard conditions. Since different types of boards can vary differently in this property as between the face and edges, usually both face and edge screw-holding strength are measured when this property is of interest. Screw-holding strength to be found among commercial particleboards ranges as follows:

	Face	*Edge*
High Density	350 to 600 lb	350 to 500 lb
Medium Density	200 to 500	150 to 400
Low Density	100 to 250	75 to 150

G. Dimensional Stability

The dimensional stability of particleboard includes its thickness swelling and shrinking and to linear dimensional changes in the plane of the board with changes of moisture content.

Thickness stability of a board is usually expressed as the thickness swelling that takes place upon soaking of test specimens in water under standard test conditions. There are two components of this swelling—that which results from the normal swelling of wood associated with hygroscopic uptake of water into wood substance, and that which results from compression recovery of wood particles that have been deformed in the board pressing operation and have retained compression stresses. The former type of swelling is reversible on reduction of moisture content whereas compression recovery is irreversible.

Thickness swelling is a general guide to thickness stability and resistance, for example, to edge swelling when particleboard is exposed to edge wetting in joints with other materials that do not swell or swell differently. It is also a general indicator of surface stability or resistance of the particleboard surface to roughening when exposed to unfavorable moisture conditions. Thickness swelling also has value as an indication of the general water resistance of a board—the resistance to deterioration of mechanical properties on excessive exposure to moisture. Water absorption upon soaking is often also specified in assessing board quality. This has some correlation with thickness swelling and general water resistance characteristics of a board, but this correlation is very limited. A water-absorption specification is largely superfluous if thickness swelling has been specified.

Thickness swelling characteristics of commercial particleboard run mainly within the following limits:

High Density	4 to 25%
Medium Density	3 to 25
Low Density	2 to 15

Linear stability is one of the more important physical properties of particleboard because it relates not only to the constancy of the length and width dimensions of particleboard in use but also to the resistance of particleboard to warping in use. Particleboard is often exposed to conditions that promote warping. These may be differences of temperature and humidity or different kinds of finishes or overlaying materials on the two faces of the board. In conditions favoring warping, the greater the linear stability of the board, the lower the the risk of warping.

Linear stability is determined by exposing board test specimens to one level of relative humidity (50%) until they reach equilibrium, then exposing them to a higher relative humidity level (90%) until a second equilibrium is reached, and measuring the length increase between the two equilibria. Although linear stability is one of the most important properties of particleboard, it is not routinely tested because the test is a long and cumbersome procedure. Research has been conducted on a relatively fast test known as the vacuum-pressure-soak test to establish correlations with the standard linear expansion test. The test is used in some plants.

The range of linear expansion values found in commercial board is as follows:

High Density	0.15 to 0.85%
Medium Density	0.15 to 0.60
Low Density	0.15 to 0.30

H. Working Characteristics

Since much particleboard is a raw material for manufacturing operations, its working or machining characteristics are often of importance. There are no standard tests by which these characteristics can be quantitatively determined. Nevertheless, users evaluating different boards for a specific use requiring appreciable machining often subject samples of the various boards to appropriate machining tests—sawing, planing, drilling, routing, shaping, etc.—and note smoothness of the surfaces produced (particularly on edge surfaces), sharpness of corners, and absence of chip-out or torn grain.

Coating and gluing characteristics may also be subject to empirical tests to evaluate boards for a particular end use.

One working property of particleboard relates to its capability of being overlaid with "one-shot laminates." This type of laminate involves laying up resin-impregnated decorative papers with particleboard cores and consolidating the overlay directly onto the core. This process is distinguished from consolidating the plastic laminate overlay in a prior operation and then simply gluing the preconsolidated laminate to the core. Because the consolidation of the decorative laminate normally necessitates considerably higher pressures and temperatures than a simple gluing operation, the core particleboard requires high compressive strength and resistance to heat. Particleboard of higher densities, bonded with phenolic resin, is usually specified for this purpose.

I. Durability and Aging Characteristics

Under appropriate conditions wood is basically a highly durable material, and the bonding agents used in commercial particleboard—urea-formaldehyde and phenol-formaldehyde resins—are themselves also durable. Therefore particleboard utilized in interior applications is possessed of high durability, and this is borne out in material that has been in service for more than twenty years. The more severe conditions represented by exterior exposure, particularly where the stuctural properties and surface integrity of the board must be assured for many decades, present a special problem to manufacturers and users of particleboard. The prospects of increasing exterior use of particleboard in applications where structural requirements must be met necessitates determination of properties of particleboard under conditions that are both severe and of long duration. The most generally recognized system of tests was one developed by the National Bureau of Standards in 1926 and designated as Accelerated Aging (see Section VI).

The test requires repeated cycles of soaking in water, spraying with steam, freezing, drying at high temperature, spraying with steam, and redrying at high temperature.

IV. Variation of Properties Through Manufacturing Controls

The early particleboards were produced by crude methods requiring much hand labor and allowing little control of processing variables. However, the industry has now advanced to a high degree of technical sophistication such that slight adjustments in the process can be made

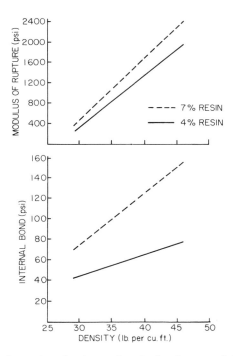

FIG. 5. Effect of varying density and resin level on modulus of rupture and internal bond properties of shavingsboard.

to provide wide variations in board properties. A considerable amount of research has been performed to study the effect of many of these manufacturing variables, some of which interact in complex ways to modify board properties. For simplicity, these factors will be discussed separately in the following paragraphs.

A. DENSITY AND DENSITY PROFILE

As noted under Properties, board densities range from below 37 lb/cu ft to over 50 lb/cu ft. Density has a pronounced effect on board properties; and, since it is so easily controlled in the process, it is one of the most important factors to be recognized in the realm of particleboard. Changes in modulus of rupture and internal bond may be accomplished as shown in Fig. 5 by varying density and resin level (effect of resin will be discussed later). Changes in density can be achieved quite easily in the process, with accompanying changes in board properties.

However, there is also a density profile or a "layered density" in particleboards. Because of the rheology involved in the manufacturing process, even if the same particle is used throughout, there is a definite pattern of density variation throughout the thickness of any board. The density profile is primarily a function of the press cycle, and its development will be discussed later. In modern particleboards, the density profile of a particleboard of 46 lb/cu ft can range from about 62 lb/cu ft at the surface to 40 lb/cu ft in the core, a dramatic range of values as

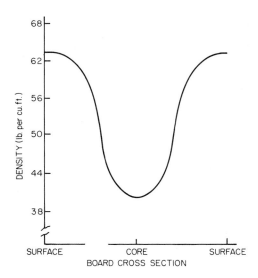

FIG. 6. A typical density profile of a particleboard.

illustrated in Fig. 6. Since density is an indicator of board properties, the prediction can easily be made that this board has enhanced surface properties and diminished core properties. With a board of homogeneous particles evenly spread with resin, this is probably an accurate statement; however, the other variations to be discussed in this section can alter this situation to some extent.

Virtually all particleboard now produced has a density profile somewhat similar to the one just described.

B. Press Time and Cycle

All boards must be hot pressed for a period of time to cure the resin. The press cycle normally followed includes an initial high pressure period to consolidate the mats to the proper thickness, followed by a curing period in which the pressure on the board is gradually reduced.

Pressure reduction is necessary as the board cures, because as the resin bonds are established to hold the particles in their compressed state, the resistance of the board to the consolidating pressure is decreased. Otherwise, if the high initial pressure were maintained, the boards would be pressed thinner than desired. If stops were used to control board caliper, the high pressure would build up on the stops as the mat resistance decreased and this pressure probably would deform the press plates.

Ten years ago, a typical press time for ⅜-inch urea-bonded board was about 10 minutes. Advances in resin technology have drastically reduced press times for ⅜-inch board to about 4 minutes, and by one method to as low as 2 minutes. The total press time required is roughly proportional to the board thickness, as sufficient heat to cure the core must be conducted from the surfaces in contact with the press hot plates, which are at about 325°F. Approximately 210°F must be reached in the core to accomplish an adequate cure in urea-bonded boards.

The press cycle is characterized in modern plants by a fast "closing time" of about 2 minutes. Closing time, in the initial high pressure period, is the time interval between loading the mats into the press and consolidating the mats to final thicknesses. About 500 psi on the mat is required for this relatively fast closing. As the mat cures into the final board and loses its tendency to spring back to its original thickness, the pressure reduction phase of the cycle takes place, until, at the completion of the press time, pressure is below 100 psi.

The character of the previously mentioned density profile shown in Fig. 6 is primarily a function of press closing time. The faster the closing rate, the higher the surface density and the lower the core density. This is because the heat, acting first on the surface layers in contact with the hot press plates, induces a disproportionate amount of the total consolidation to occur at this point. As soon as final board thickness is reached, penetration of the heat into the core layer occurs after pressure is reduced. Hence, the core structure is made rigid by the curing action of the resin while in a less compacted state. Moisture present in the surface layers also responds to the heat to enhance compaction differentially in the same time sequence. If the time interval to reach final board thickness is increased, the combined effect of heat, pressure, and moisture promotes more nearly equal compaction inward toward the center of the board. However, if this interval is too great, the surfaces will be of low density and low strength because the curing action of the resin will precede the compaction action.

Thus, by varying the press closing time, the density profile can be changed, with concomitant variations in board properties. Essentially, fast closing times yield high bending and low internal bond strengths, whereas, in general, a reversal in these properties occurs when slow closing times are used. For economic reasons, the faster closing speed is usually preferred because it enhances the production capacity of the plant. Early particleboard presses closed much more slowly, but the slow close resulted in soft surfaces that had to be sanded off, which wasted much valuable material. In fact, some of these presses took longer to close than the total press time now possible.

Many of the early multiopening particleboard presses were modified plywood presses that closed from the bottom up. Thus, the lower openings were closed first. This produced a set of boards from each press load that had varying density profiles. The massive multiopening presses now available have ingenious simultaneous closing devices that close all press openings at the same rate. Consequently, density profiles are kept constant for all boards.

The new press developed for curing the medium-density fiberboards employs a combination of heated plates and dielectric heating, or as it is commonly called, radio frequency (rf) heating. This gives an approximately simultaneous curing of the board surfaces and core that is unlike the curing effected in a standard press. Thus, a relatively balanced density profile is developed in this board. Further developments with rf heating could provide even more sophisticated presses for the manufacture of particleboard.

C. Effect of Moisture

Moisture contents of the mats have to be kept below 12%, as higher moisture levels interfere with proper resin bonding. In addition, the moisture content must be constant throughout the mat or, if a differential in moisture content by layer is preferred, it must be precisely controlled. Precise control is now absolutely necessary to avoid "blowing" the board. A blow is a steam pocket formed during pressing, usually because of excess moisture. The vapor pressure of the steam exceeds the internal bond strength of the "hot" board, and when the press is opened, the steam pressure blows the board apart.

The best example of a controlled differential moisture level in the mat by layer, and its effect on particleboard properties is the so-called "steam shock" treatment. Wood itself has good insulating properties, which makes it difficult for heat to transfer rapidly to the mat core while it is being pressed into a board. The steam shock treatment is used to speed up the heat transfer. In this treatment, a light spray of water is applied to the mat surfaces prior to its entering the press, and when the hot plates close on the water-treated mats, the water turns to steam. The steam rapidly passes through the minute interparticle passageways to the core and quickly heats it to the curing temperature of the resin. The steam shock treatment has the further advantage of plasticizing the surface particles, which assists in compressing the surface layers into hard, smooth faces on the board.

Some plants with fast press times keep the mat moisture content down to 7 or 8% to avoid blowing problems. Other plants keep the moisture content low because of gases that can form from extractives in the wood and help to build up the vapor pressure within the board. A case in point is western red cedar, which has a high extractives content. A decrease in moisture content, however, reduces the plasticizing effect of the steam on the wood particles. Consequently, higher pressures are usually needed to consolidate such mats into boards.

D. TYPE AND AMOUNT OF RESIN

It has been stated earlier that most particleboard is bonded with urea-formaldehyde resin and that phenol-formaldehyde resin is used for the limited production of board that will be subjected to exterior exposures or certain secondary gluing operations. Other resins or glues have not been used because of high cost or poor performance characteristics. Advances in resin technology may, at any time, develop a new resin that will preclude the use of ureas or phenolics.

The resin usually is specifically formulated for each plant and is generally supplied by chemical companies although some is formulated in the plant. The amount varies with the product being manufactured. Higher quality board such as "core stock" may average about 8% in resin content, whereas floor underlayment may contain approximately 5%. Increased resin content, as would be expected, improves board properties. In the experiment summarized in Fig. 5, referred to earlier, both resin level and board density were varied in shavings boards. Increasing both these factors provided better modulus of rupture and internal bond values, although it is noticeable that internal bond benefited the most from both increased resin and density. In general, the internal bond value can be used as an indicator of most board properties.

When improved board properties are wanted, it is sometimes less expensive to increase density rather than resin content because of material cost factors. More of the resin is effectively used in the board of higher density because of more intimate contact between particles. Resin is wasted in lower density boards when located in interparticle voids. Moreover, the higher density board has a better appearance because of less porosity, especially on the edges.

The resin is applied as tiny droplets by the spraying system in the blender, and usually the particles are held together with "spot welds" in the final board. This process is unlike most wood gluing, which is performed with continuous films of glue over all of the surfaces being mated. It has been found that varying the size of the resin droplets affects board properties. Droplets averaging about 30 microns in diameter usually provide optimum bonding conditions. Therefore, precise control of the resin spraying system is extremely important if the plant is to function economically.

Boards have a layered density profile as mentioned. As will be discussed under Particle Geometry, much of the production also purposely layers small particles on the surfaces and coarser particles in the core. Edge views of two boards, one layered and one using the same size particle throughout, are shown in Fig. 7, to illustrate the difference between layered and homogeneous boards. Resin levels of the surface and core particles can be varied to develop the board properties desired. The smaller particles have a greater surface area per unit weight and therefore logically should require more resin than the coarser particles. However, it has been found that on a surface-area basis, the coarsest particles, located in the low-density core, require more resin than the fine particles used for surfaces, if the optimum balance of board properties is to be achieved. The finer particles used for the surfaces make more effective use of the resin applied because of the higher densities of the surface layers in most commercial boards.

Two systems are used for blending resin with the furnish. The first treats all particles simultaneously; the second treats surface and core material independently. In practice, blenders treating all particles simultaneously can apply as much as 20% resin on the fine particles and as little as 1% on the coarse particles while maintaining an average resin level of 8%.

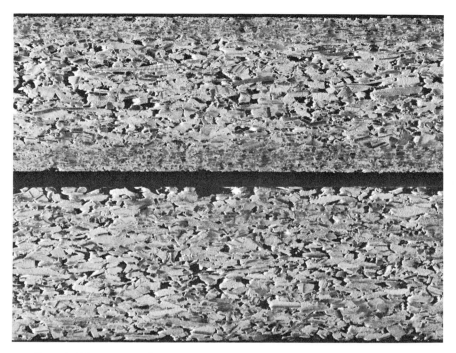

Fig. 7. Edge views of two particleboards. The top view shows a board with pronounced high-density surfaces composed of fine particles and a low-density core of coarse particles. The bottom view is a board of homogeneous particles and a relatively uniform density profile.

Refinements in this blender design, however, have made it possible to exercise much greater control over the resin distribution. It is also possible to apply a less advanced resin, requiring a longer cure time, to the surface particles. This procedure results in a somewhat equalized cure time for the surfaces and core of the board pressing and assists in preventing precure of the surfaces, which yields soft or "punky" faces.

The other system, treating surface and core particles separately, perhaps provides the greatest flexibility in resin application, as precise control can be exercised over the resin characteristics and resin level of each layer in the board.

The first large particleboard plants were built in the Douglas-fir region of the Western United States. Douglas-fir, by chance, has a natural acidity that quickly catalyzes urea-formaldehyde resin. Consequently, bothersome resin problems attributed to wood pH did not exist, making it possible to apply the resin "neat" as received, without any need to catalyze. Plants based on other tree species have not been as fortunate, and catalysts have had to be used to insure fast press times. This gives some control problems as greater care must be exercised when using catalyzed resins that can set up in the blending system, particularly if there is an equipment breakdown. However, the catalyzed resin systems have been developed to such an extent that most species can be used in the modern, fast board-processing systems.

A particular disadvantage in the use of urea resins in combination with fast press times is a residual formaldehyde odor in the finished boards. Normally, a chemical scavenger is incorporated into the resin to alleviate the odor problem. Some odor problems persist occasionally, especially if the board is not sealed or overlaid. Whenever particleboard is exposed in tight quarters, such as in a closet, it is always good practice to apply a seal coat of finish. Formaldehyde odor normally is not noticeable in larger rooms.

E. PARTICLE GEOMETRY

It has been stated that boards layered by particle size are the most commonly produced. Two types of layered board are made—one multi-layered, the other having three distinct layers. These two board types are the dominant ones in the industry and are based for the most part on planer shavings. A typical range of particle sizes found in a board is shown in Fig. 8. The larger particles will pass though a No. 6 screen

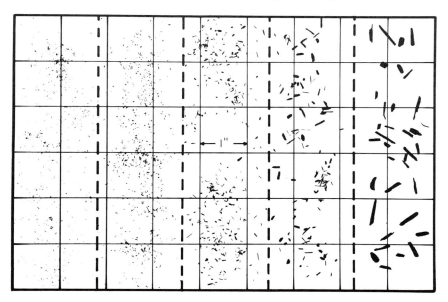

Fig. 8. Typical particles used in a layered particleboard.

(0.132-inch openings); the finer particles are dustlike. The multilayer board is formed with a gradation in particle size from finest on the surfaces to coarsest in the core. The two coarsest particle sizes shown usually make up about 60% of the board. For a three-layer board, the three classifications of finer particles are mixed for surface layers and the two coarser fractions are mixed for the core layer. The two surface layers each comprise about 20% of the board, and the core contains the remaining 60% of material.

In addition, some homogeneous board is produced in which all particles are randomly mixed. These are normally used for applications where the smooth surfaces developed by fine particles are not necessary.

It has been noted before that the predominant particle used for particleboard in the United States is the shaving generated by planing lumber. Such shavings vary widely in geometry depending upon the species and whether the lumber is planed green or dry. Some species yield a thin, flake-like shaving that can contribute to bending strength in the board. Others yield block-shaped particles that intermesh well in the board and provide superior internal bonding. It is possible to place the different shavings in the board in such a way as to take advantage of their varied particle geometries in producing the multilayer or three-layer type of board.

Plants operating on a single tree species can usually employ simple screening to sort out the particles they wish to process. Oversize ones are then passed through some type of mill, reduced in size and fed back into the system. As the industry has evolved, it has been necessary, at times, to use smaller particles to provide boards with fine surfaces and tight edges. Thus, many of the larger shavings must be milled to a smaller size to fit the smaller particle size limits established. Consequently, inherent particle geometry differences between shavings from different species can be minimized by the reduction in size.

For surface layers, it is also possible to use small fibers, which contribute to bending strength with their superior length-to-thickness ratio. In other words, the long, threadlike fibers are a superior particle for faces because bending strength is improved while a smooth, hard board surface is retained. Again, this is a layered particleboard with coarse particles in the core.

Over a decade ago, the popular belief was that flakes were such a superior particle for board that their use would overshadow any other particle. However, the precision distribution of the varied sizes of planer shavings within the board, along with the technological advances described under resin distribution and press cycles, has made it possible to produce quality board of planer shavings. This development provided a tremendous economic breakthrough because shavings prepared for board processing at the blender cost about $7 per ton, whereas flakes cost about $26 per ton.

The major particle generated especially for particleboard is the flake. Flakes can be cut with the length, thickness, and width desired for the product. Different sizes can be cut for the various layers so that properly "designed" flakes are preferentially placed within the board to develop the optimum board properties required. For example, thick flakes can be in the core and very thin ones on the surface. This yields a board with high bending strength, good internal bond and a relatively smooth surface. It is also possible to combine flakes with shavings or fiber. All of these particle combinations along with varying resin distributions and density profiles can provide a number of boards with different physical properties. In effect, varying these factors can yield many different boards essentially from the same materials.

Flakes, while expensive to produce, are the superior particle for boards planned for structural use. It is possible to cut long, thin flakes and orient them parallel to each other in the board. This produces a structural board with high bending properties. In fact, it is possible to make such an oriented board out of a low-density wood, such as aspen, so that it has greater bending strength than the original wood. Production of particleboard for structural uses is just beginning. Extensive research is needed to assist this advancement of the industry.

Sawdust is being used now in small amounts in the particle mix. Sawdust, because of the way it is generated, has relatively short fiber length and many of the particles are blocky in shape, containing incipient fractures. However, as the use of smaller particles in the board furnish increases, sawdust can be used more extensively. The development of the pressurized refiner is also making it possible to generate a fibrous particle from sawdust that is well suited for part of the furnish.

F. Species of Wood

The primary effect of species upon board properties is density. A low-density wood can be used to make board of a much lower density than a high-density wood. Some hardwoods in the natural state are heavier than medium-density particleboard. Since some pressure is required in consolidating the mat into a board to effect adequate resin bonding between particles, it is impossible to produce a low- or medium-density board from a high-density wood. Thus, conventional particleboard is limited to species approximately 30 lb/cu ft in density or lower.

Woods of higher density naturally can be used for high-density boards and it is possible to add some high-density woods to a medium-density particleboard furnish in controlled amounts.

A problem with a wide range of species in the furnish, especially in undetermined amounts, is loss of control of the furnish pH. This makes it difficult to formulate the resin in balance with the furnish pH. In practice, it is desirable to control the mix of species so that the resin will be able to cure properly and at a constant rate in every board.

A sensible arrangement when dealing with a number of species is to segregate them so that inherent particle geometries, density variations, and pH problems can be isolated. Then, production advantages can be taken of the superior qualities present in the species and the harmful qualities can be either treated or worked into the process in such a way that minimum damage is done to board properties.

G. Layering Within the Board

As has been described, many types of layering can take place within the board. Variations in density and density profile, resin level and distribution, particle size and geometry, and species all must be taken into account. If all of these factors and their interactions are controlled, boards meeting various production standards can be produced from a wide variety of materials. Neglecting to consider and control these factors will instead cause wild variations in board properties. Consequently, quality control measures are of extreme importance if a particleboard plant is to manufacture a product that has consistent physical properties.

H. Special Additives for Fire and Decay Resistance

Fire-retardant and decay-resistant particleboards are being made on a limited basis. Extensive development both in basic research and marketing is necessary before these products become major items. However, the small particles comprising the board furnish are ideal for treatment with various additives prior to board pressing. Solid wood is very difficult and expensive to pressure treat for fire retardancy, decay resistance, and dimensional stability. Perhaps particleboard is the most fertile field for the development of the wood product that has these qualities. It might be said that this would be the perfect wood product.

V. Applications of Particleboard

The two major applications of particleboard are manufactured articles, such as furniture and casegoods (core stock), and building construction.

A. Core Stock

In the manufacture of furniture and casegoods, particleboard is employed to best advantage in those parts having large, flat surfaces. Most particleboard in this area of application is overlaid either with decorative wood veneers or plastic laminates or films, as shown in Fig. 9. Hence, much of the particleboard produced is designated as "core stock." Examples of uses in this category are table tops (dining room, dinette, coffee, end library, game, writing, and work tables), buffets, credenzas, office desks, school desks, chair backs, chair seats, church pews, chests of drawers, bureaus, counter tops, cabinets (especially doors and drawer fronts, but also sides, tops, bottoms, and backs), and store fixtures. In the core stock application, particleboard has extensively supplanted plywood and lumber. Presently, approximately 60% of the production is core stock.

In industrial operations related to production of these items, special attention is necessary to use of proper glues and gluing methods in ap-

FIG. 9. Particleboards to which typical overlays have been applied.

plication of the overlaying material. One broad requirement is to employ as near as feasible a "balanced" construction to insure against the possibility of warp. A balanced construction requires surfacing both faces of the particleboard with materials (whether coating or overlay) having the same or similar properties with respect to permeability to moisture and tendency to shrink or swell with ambient temperature and relative humidity.

Treatment of the particleboard edges when exposed to view is also a matter of particular concern. The most common approach is to band the edges with lumber, wood molding, veneer, plastic laminate, or plastic film. The use of an edge filler in combination with an edge coating matching the face treatment is also common.

B. Building Construction

In building construction, the principal use of particleboard is as underlayment for flexible flooring materials, in the form of either tile or sheet goods. Smooth, uniform surface available in combination with hardness and dent resistance makes particleboard serve especially well in this application. Other building material applications are wall panels, room dividers, wainscots, door cores, mobile-home subfloors, shelving, flooring, soffits, and exterior siding.

The growing use of particleboard as a core material for solid-core flush doors is recognized in the National Woodwork Manufacturers Association Standard For Hardwood Veneered, Including Hardboard and Plastic Faced, Flush Doors, NWNA I.S. 1–66. Particleboard is listed here along with the long-established solid wood and mineral cores. Similarly, the use of particleboard as a core for hardwood plywood, along with veneer, solid wood, and hardboard, has been recently recognized in a commercial standard for hardwood plywood (U. S. Dept. of Commerce Commercial Standard CS35-61).

The first major use of particleboard as a structural material has developed recently. Boards ⅝ inch by 8 by 24 feet are used for the floors of mobile homes. These panels are nail-glued to the structural framework and simultaneously serve as the floor diaphram and the support for plastic floor coverings or carpet.

In addition to the above-mentioned major areas of use, particleboard has a wide variety of miscellaneous uses. Examples are toys, signs, musical instruments, coffins, and printing blocks. One large potential area of use that has not at present been realized is in packaging.

VI. Product Standards and Trade Associations

Because properties and quality of particleboard can vary within an extremely wide range depending on the conditions of manufacture, particleboard manufacturers early in the development of the industry found it desirable to set up industry standards to protect both the consumer and the responsible manufacturer. The standards were promulgated through the Office of Product Standards in the National Bureau of Standards of the U. S. Department of Commerce. The particleboard standard currently in effect is designated as Commercial Standard CS236-66, "Mat-Formed Wood Particleboard." The coverage of this standard is best described by quoting the section giving the scope of the standard:

"2.1 Scope—This Commercial Standard covers two types of mat-formed wood particleboard; one for interior applications and one for certain exterior applications in addition to interior applications. Each type is further divided into several density grades which are subdivided into strength classifications. It is intended that the applications of the products will be consistent with the properties of the respective grades and strength classifications described. Also included are definitions, dimensional tolerances, test methods, inspection practices, and method of marking and certification to identify products that comply with all requirements of this Standard."

The heart of this standard is the table giving property requirements of the different types, density grades, and classes of particleboard. This table (Table II) is herewith reproduced in its entirety. The values for properties given in the table are designed to serve as lower limits for properties specifications, and most manufacturers publish company specifications that equal or exceed those of the Commercial Standard.

Since virtually all extruded boards are produced for internal consumption by captive plants, there are no commercial standards covering extruded boards.

In addition to the foregoing standard there is extant a particleboard standard covering exterior particleboard, promulgated by the West Coast Particleboard Manufacturers Association. The main feature of this standard is the specification of the use of the Accelerated Aging Test and the degree of retention of various physical properties upon completion of the test.

The methods of test utilized in the various product standards were developed through activities of the American Association for Testing and Materials. The applicable standard is identified as ASTM D1037-64, "Standard Methods of Evaluating the Properties of Wood-Base Fiber and Particle Panel Materials." This standard describes a series of physical and mechanical tests listed as follows:

Size and Appearance of Boards
Strength Properties:
  Static Bending
  Tensile Strength Parallel to Surface

TABLE II

MINIMUM PROPERTY REQUIREMENTS FOR MAT-FORMED PARTICLEBOARD, COMMERCIAL STANDARD[a]

Type (use)	Density (grade) (min. avg.)	Class[b]	Modulus of rupture (min. avg.), psi	Modulus of elasticity (min. avg.), psi	Internal bond (min. avg.), psi	Linear expansion (max. avg.), %	Screw holding Face (min. avg.), lb	Screw holding Edge (min. avg.), lb
1[c]								
	A (High density, 50 lb/cu ft and over)	1	2400	350,000	200	0.55	450	—
		2	3400	350,000	140	0.55	—	—
	B (Medium density, between 37 and 50 lb/cu ft)	1	1600	250,000	70	0.35	225	160
		2	2400	400,000	60	0.30	225	200
	C (Low density, 37 lb/cu ft and under)	1	800	150,000	20	0.30	125	—
		2	1400	250,000	30	0.30	175	—
2[d]								
	A (High density, 50 lb/cu ft and over)	1	2400	350,000	125	0.55	450	—
		2	3400	500,000	400	0.55	500	350
	B (Medium density, between 37 and 50 lb/cu ft)	1	1800	250,000	65	0.35	225	160
		2	2500	450,000	60	0.25	250	200

[a] From Commercial Standard CS236-66, U. S. Dept. of Commerce, Office of Product Standards.
[b] Strength classifications based on properties of panels currently produced.
[c] Mat-formed particleboard (generally made with urea-formaldehyde resin binders) suitable for interior applications.
[d] Mat-formed particleboard made with durable and highly moisture- and heat-resistant binders (generally phenolic resins, suitable for interior and certain exterior applications when so labeled.

Tensile Strength Perpendicular to Surface
Compression Strength Parallel to Surface
Fastener Holding:
  Lateral Nail Resistance Test
  Nail Withdrawal Test
  Nail-Head Pull-Through Test
  Direct Screw Withdrawal Test
Hardness Test
Shear Strength in the Plane of the Board
Glue Line Shear Test (Block Type)
Falling Ball Impact Test
Abrasion Test
Moisture Tests:
  Water Absorption and Thickness Swelling
  Linear Variation with Change in Moisture Content
  Edge Thickness Swelling by the Disk Method
Accelerated Aging
Cupping and Twisting
Moisture Content and Specific Gravity

The General Services Administration of the U. S. Government has promulgated a particleboard specification LLL-B-800 a (May 15, 1965), Federal Specification, "Building Board (Wood Particleboard), Hard Pressed, Vegetable Fiber." This standard is used in materials procurement activities of various agencies of the federal government. It incorporates the Commercial Standard CS236-66 by reference and in addition covers other specifications, particularly with respect to packaging for shipment and inspection procedures. In addition to this general specification, particleboard is covered in numerous other Federal and Military Specifications applicable in government purchases, both nonmilitary and military, of household, office, and institutional furniture, and casegoods.

The specific use of particleboard in building construction is mainly covered by standards specified by the Federal Housing Administration in the U. S. Dept. of Housing and Urban Development. Although these standards are mandatory only in residential construction covered by FHA insured mortgages, they have much broader influence because they are used to a wide extent in other residential contsruction in which FHA is not involved. The major use of particleboard is as flooring underlayment but also includes sink and cabinet tops and cabinet and wardrobe doors. These uses are covered in FHA "Minimum Property Standards for One and Two Living Units" (MPS No. 300), and "Minimum Property Standards for Multi-Family Housing" (MPS No. 2600). These standards specify the conditions of use of particleboard in flooring underlayment application and incorporate by reference FHA Use of Materials Bulletin No. UM-28a titled, "Mat-Formed Wood Particle Board for Floor Underlayment," which in turn references the National Particleboard Association's "Physical Properties Standards for Mat-Formed Particleboard for Floor Underlayment." The use of particleboard in kitchen counter tops and sink tops in residential construction is also provided for in these standards.

In view of the beginning penetration of particleboard in other house construction applications, FHA has also provided for the use of exterior particleboard in siding and other nonstructural applications through a Use of Materials Bulletin, No. UM-32, titled, "Mat-Formed Particleboard for Exterior Use."

Other government specifications regulating the use of particleboard in residential construction are the Department of Defense "Guide Specification for Military Family Housing" (DOD 4270.21—SPEC) covering floor underlayment, and the Department of Housing and Urban Development Housing Assistance Authority "Guide Specification for Low Rent Housing" (Bulletin LR-13) covering floor underlayment and sink-top corestock.

The American particleboard industry is represented by the National Particleboard Association with offices located at 711 14th Street, N. W., Washington, D. C. 20005. The Association members account for about 80% of the production in the United States. Of particular interest to the user and specifier of particleboard is the Association's Technical Committee, which is charged with the responsibility of developing industry product standards. Another major responsibility of the Committee is to encourage proper use of particleboard in its expanding and diversifying applications.

VII. Future Outlook

The particleboard industry, although it has recently passed the one billion square feet annual production mark, is still in its infancy. The extreme flexibility of the production operation with respect to the type of raw material, the control of particle size and shape, the kind and amount of binders and other additives, and the pressing conditions make possible a wide range of products to suit various requirements. The principal end uses of particleboard up to the present can be classified as being in furniture and cabinetry and miscellaneous industrial purposes. Vigorous growth in this area of utilization can be expected to continue. However, the largest visible potential for expansion lies in the building construction field.

One particular natural advantage enjoyed in particleboard manufacturing is the ease with which additives may be incorporated into the board to produce special properties. Prominent among such additives are preservatives against insect attack and decay. Fire-retardant agents also give promise of opening large new areas of application for particleboard in the building materials field.

The birth and initial growth of this newest of the forest products industries was largely fostered by heavy investments in research and development. The continuing efforts in this field promise large increases in volume and proliferation of particleboard products.

SELECTED BIBLIOGRAPHY

Books

1. L. E. Akers, "Particle Board and Hardboard." Pergamon, Oxford, 1966.
2. H. J. Deppe and K. Ernst, "Technologie der Spanplatten." Zentralblatt, Stuttgart, 1965.
3. E. S. Johnson, ed., "Wood Particle Board Handbook." North Carolina State College, Raleigh, North Carolina, 1956.
4. F. Kollmann, "Holzspanwerkstoffe." Springer, Berlin, 1966.
5. T. M. Maloney, ed., "Proceedings of First Washington State University Symposium on Particleboard." WSU, Pullman, Washington, 1967.
6. T. M. Maloney, ed., "Proceedings of Second Washington State University Symposium on Particleboard." WSU, Pullman, Washington, 1968.
7. T. M. Maloney, ed., "Proceedings of Third Washington State University Symposium on Particleboard." WSU, Pullman, Washington, 1969.
8. L. Mitlin, ed., "Particleboard Manufacture and Application." Pressmedia Ltd., Kent, England, 1968.
9. "Fibreboard and Particle Board." Food and Agriculture Organization of the United Nations, Vol. 1-6, Rome, 1958.

Journal Articles

1. J. Brumbaugh, Effect of flake dimensions on properties of particle boards. *Forest Prod. J.* **10** (5), 243–246 (1960).
2. C. H. Burrows, Some factors affecting resin efficiency in flake board. *Forest Prod. J.* **11** (1), 27–33 (1961).
3. M. N. Carroll, Whole wood and mixed species as raw material for particleboard, *Washington State Univ. Bull.* **274**, (1963).
4. M. Carroll and D. McVey, An analysis of resin efficiency in particleboard. *Forest Prod. J.* **12** (7), 305–310 (1962).
5. M. I. Chanyshev, "Steam shock" in chipboard manufacture. *Ind. Wood Proc.* **5**, 21–23 (1960).
6. F. Fahrni, Automation in chipboard plants. *Holz Roh Werkst.* **18** (1), 15–19 (1960).
7. B. G. Heebink and R. A. Hann, How wax and particle shape affect stability and strength of oak particle boards. *Forest Prod. J.* **9** (7), 197–203 (1959).
8. W. Klauditz, The development and the position of particleboard manufacture from 1955–1961. *Holz Roh Werkst.* **20** (1), 1–12 (1962).
9. W. Klauditz, Manufacture and properties of particleboard with oriented strength. *Holz Roh Werkst.* **18** (10), 377–385 (1960).
10. F. Kollmann, Effect of moisture differences in chips, before pressing on the properties of chipboard. *Holz Roh Werkst.* **15** (1), 35–44 (1957).

38 THOMAS M. MALONEY AND ARTHUR L. MOTTET

11. W. F. Lehmann, Improved particleboard through better resin efficiency. *Forest Prod. J.* **15** (4), 155–161 (1965).
12. G. G. Marra, Particleboards . . . their classification and composition. *Forest Prod. J.* **8** (12), 11A–16A (1958).
13. A. L. Mottet, Flakeboard vs. chipboard. *Lumberman* **87** (7), 42–43, 61 (1960).
14. C. R. Morshauser, How to specify particleboard. *Woodworking Digest* **71** (9), 30–32 (1969).
15. P. W. Post, Relationship of flake size and resin content to mechanical and dimensional properties of flake board. *Forest Prod. J.* **11** (1), 34–37 (1961).
16. H. F. Schwiertz, Particleboard forming and pressing—with and without cauls. *Board Mfr.* **11** (8), 83–87 (1968).
17. M. D. Strickler, Effect of press cycles and moisture content on properties of Douglas-fir flakeboard. *Forest Prod. J.* **9** (7), 203–215 (1959).
18. O. Suchsland, An analysis of the particleboard process. *Quart. Bull. Mich. Agr. Expt. Sta.* **42** (2), 350–372 (1959).
19. D. H. Turner, Effect of particle size and shape on strength and dimensional stability of resin-bonded wood-particle panels. *Forest Prod. J.* **4** (5), 210–223 (1954).
20. I. Wentworth, Caulless process for making particleboard. *Forest Prod. J.* **18** (1), 12–13 (1968).
21. M. G. Wright and R. B. Phelps, Particleboard, insulation board and hardboard. Industry Trends 1956–1966, *U. S. Forest Serv. Res. Paper* **WO-5** (1967).

ACOUSTICAL MATERIALS

Lyle F. Yerges

L. F. Yerges, Consulting Engineers, Downers Grove, Illinois

I. Introduction

A. Scope

It is probably unfortunate that the term "acoustical materials" has become a part of our technical vocabulary, since this suggests a family of unique, specific materials with unique properties. As a result, engineers, scientists, and product designers often tend to think of sound control and vibration control as highly specific subjects or as something to be added to or imposed upon a completed design (or as an afterthought when a completed design results in an unacceptably noisy or vibratory product or environment). The acoustical characteristics of all materials are as basic as their density, elasticity, or hardness. In fact, the acoustical characteristics of materials are directly related to the basic physical properties of the materials; this should be understood and taken into account when a material is being considered during design.

All materials are "acoustical materials" in the strict sense of the term. They absorb, reflect, or radiate sound, and they damp vibrations. However, this chapter deals with those materials whose physical characteristics particularly fit them for sound and vibration control.

During the past few decades, a whole fabric of myths, partial truths, preconceptions, and misconceptions about acoustics and acoustical materials has developed, even among the technically trained. A discussion of the simple, basic principles will be undertaken here to give practicing professionals and others a fundamental knowledge of the subject.

No attempt will be made to examine closely the theory or the mathematics of most phases of acoustics, since this information is readily available in many excellent textbooks and handbooks. Neither will there be an extensive listing of data which become quickly obsolete. The performance data and characteristics that will permit the designer to choose materials for actual use, to specify, to evaluate, and to design will be covered broadly.

Acoustical materials are essentially transducers. Usually, they convert mechanical energy to thermal energy. The absorption, reflection, transmission, and radiation of acoustic energy by various materials constitutes essentially the whole field of sound and vibration control.

The measurable, predictable, and controllable characteristics of materials are the data of interest to the designer. Performance characteristics in acoustics, as in most fields are given in terms of coefficients or relative measures. Usually, absolute measures are not obtainable, nor are they particularly meaningful in themselves. What is required for most engineering purposes is a vocabulary of symbols, dimensions, quantities, or other terms to communicate the information necessary to define, evaluate, compare, and choose materials and constructions. Tests are merely standardized, repeatable means of measuring or comparing characteristics of materials.

Although technological developments tend to make particular products and materials obsolete, the basic characteristics that make materials "acoustical" are easily defined and identified. New products and materials will continue to be judged and selected according to the well-established parameters discussed in this chapter.

B. Definitions

Acoustics is the science of sound, including its production, transmission, and effects.

Sound is a vibration in an elastic medium.

Acoustic energy is the total energy of a given part of the transmitting medium, minus the energy that would exist in the same part of the medium if no sound waves were present. The transmission and dissipation of acoustic energy is of interest to nearly everyone involved in engineering and architecture.

Vibration is the oscillating motion of the media; generally the term is applied to the nonaudible acoustic phenomena as differentiated from the audible.

Hearing is the subjective response to sound, whereas *feeling* is the usual subjective response to vibration.

For a complete glossary of acoustical terms, any of several complete textbooks or handbooks on acoustics and sound control will be found useful. Definitions of specific terms relating to specific subjects will be given later in the chapter.

C. Historical Development

The art or practice of sound control is probably very ancient. Considerable acoustical knowledge is attributed to the early Greeks and Romans, although most reports of the excellence of their work are without much foundation. The use of reflectors and focusing devices is probably very old. It has been known for centuries that air-borne sound transmission through massive structures is minimal, whereas tapping or impact sounds travel readily through most rigid or hard materials. Although considerable theoretical and experimental work was done by Galileo, Newton, and Helmholz, it was Lord Rayleigh who made the most significant contributions to the theory of acoustics, near the end of the 19th century (1).

Probably the most significant early contributions to architectural acoustics were the pioneering efforts of Professor Wallace Clement Sabine from about 1895 to 1920. From his study of the acoustics of buildings at Harvard University during his tenure as a physics professor, he developed the basic understanding of the relationship between reverberation and acoustical absorption in rooms (2). Up to that time little was known about the problems of room acoustics. Wondrous acoustical properties were ascribed to wires strung about in rooms, broken glass under the floors of concert halls, and large vases set about in theaters. Wood panels, particularly very old wood, were believed to be highly beneficial to the acoustics of a space. But, like most widely accepted beliefs, most of these were demonstrated to be myths.

By manipulating the number of upholstered seat cushions in a particular room, Sabine was able to control the reverberation time of the space and as a result, the intelligibility of speech within the space. He concluded that the soft, porous cushions were absorbing some of the acoustic energy reaching them and that there was a predictable relationship between the sound within the room and the number of seat cushions in the space. From this he developed the well-known reverberation time formula, which is still the basis of much acoustical design and many test procedures.

Sound transmission through walls, floors, and ceilings was studied early in the United States by the U. S. National Bureau of Standards, particularly by Buckingham (3), and later by London, Cook, and Waterhouse. Later work, particularly that of Cremer in Germany and Kurtze and Watters in the United States (4), showed the significance of the shear wave sound transmission through panels. This work has given us a much better insight into the problems of air-borne sound transmission through building structures.

The theory of vibration isolation, and the mechanism and effect of damping are too well known to require extensive discussion, since they are dealt with in nearly every theoretical and empirical study of the dynamics of vibrating bodies. However, recent developments in materials and products are worthy of examination.

II. Acoustics

The subject of sound may be reduced to relatively simple principles that can then be related to the characteristics and use of materials in sound and vibration control. It is imperative to understand the nature of sound propagation if the principles of sound control are to be correctly understood and common misconceptions avoided.

A. Sound

Sound is a relatively simple form of energy, causing variations in pressure and alternations in direction of molecular movement within the media. Usually sound propagates as a longitudinal wave, with the direction of propagation parallel to the motion of the particles. (An important exception to this principle will be discussed in a later section.)

Sound originates with a source—energy input of some sort; travels via

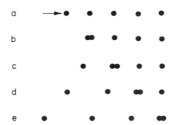

Fig. 1. Motion of molecules in elastic media as a sound wave progresses. a, Molecules at "rest," energy is applied to first; b, first molecule strikes second; c, second strikes third, first starts slowly back toward its original position; d, third strikes fourth, second starts slowly back toward its original position; first is back at original position; e, fourth strikes fifth, third starts slowly back toward its original position, second is back at original position, first is now stopped and at maximum position beyond its original position.

a path—an elastic medium of some type; and reaches a receiver—usually the human body, the receiver that is of interest to us.

B. Elastic Matter—The Sound Transmission Path

The nature of "elastic matter" explains what takes place as the sound motion occurs. Molecules in any substance are constantly moving at high high speed, the rate depending on temperature and pressure in the medium. They are striking one another, rebounding, and striking other particles. Sound motion is superimposed on this already existing motion. For the purposes of this discussion, however, we will suppose an imaginary instant in time when all particles are equally spaced, and call their position at this instant "rest."

If we were to isolate a few molecules, approximately the sequence shown in Fig. 1 would be observed. Note the motion of the first molecule in the "close-up," Fig. 2. It was displaced first to one side of its normal

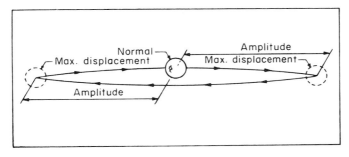

Fig. 2. Motion of a molecule during a single cycle.

rest position, then back through normal to the opposite position, and then again to normal. If this motion is regular and repetitive it is called a *vibration.*

C. The Sound Wave

Figure 3 shows what the molecules in a medium look like when a sound wave passes through it. The wavelength of the sound wave is the distance between areas of identical rarefaction or compression. As may be seen from Fig. 2, a cycle is a complete single excursion of the molecule. Frequency is the number of cycles in a given unit of time, usually cycles per second. (Cycles per second are frequently labeled hertz, abbreviated Hz.)

It is readily apparent that all of the preceding information applies equally to invisible sound vibrations and to visible or physically discernible mechanical vibrations.

The amplitude of motion is the maximum displacement, beyond its normal or rest position, of the element being considered. In most audible sounds, these excursions are very small, although low-frequency sound may cause large excursions (such as would be observed in the motion of a loud speaker cone reproducing very low-frequency sounds at audible level). In some mechanical vibrations, the amplitude of motion can be very great, particularly for very low-frequency vibration.

The important characteristics of elastic media are mass and density (remember that mass is the characteristic that imparts inertia to a substance). Elasticity and stiffness are also fundamental. In addition, plasticity and hysteresis are fundamental characteristics of media considered in acoustical design.

In a given medium, under fixed conditions, sound velocity is constant.

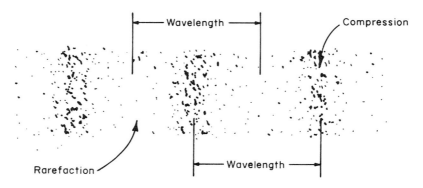

Fig. 3. Molecular motion in an elastic medium subjected to sound.

Therefore, the relationship between velocity, frequency, and wavelength can be expressed by the equation $V = FW$. Obviously, the higher the frequency, the shorter the wavelength.

Contrary to an almost universally held opinion, sound does not travel faster in more dense media. In fact, sound velocity in media is inversely proportional to the density, as can be seen from the following equation:

$$V = k(E/\rho)^{1/2} \tag{1}$$

where k is a constant, E is the modulus of elasticity, and ρ is the density.

What is usually overlooked is the modulus of elasticity of the medium. Usually, very dense materials have very much higher moduli of elasticity than do less dense media. As a result, naive and uncritical examinations of the velocity of sound in various media lead observers to precisely the wrong conclusion. It would seem logical to almost any observer, however, that the progression of the impulse through a medium would be much less restricted in less dense media, and examination of the previous sketches would bear this out. This concept is fundamental to a true understanding of transmission or isolation by acoustic media. Sound velocities in representative media are given in Table I.

TABLE I
VELOCITY OF SOUND IN VARIOUS MEDIA

Material	Approximate sound velocity, ft/sec
Air	1100
Wood	11000
Water	4500
Aluminum	16000
Steel	16000
Lead	4000

The energy concepts of interest to the designer or the user of acoustical materials are kinetic energy, potential energy, energy storage, and energy conversion. Since acoustic phenomena are equivalent to alternating electric current, the concepts of resistance, capacitance, impedance, intensity, and pressure are equally applicable. Electrical and mechanical analogies are frequently used in discussing acoustical concepts.

The most fundamental concept, however, remains the simple and well-known equation, Force = Mass $\times$ Acceleration, or $F = MA$. If this were remembered and understood, any intelligent designer could intuitively find his way through many acoustical problems. All materials possess mass; this mass, when moved, particularly when moved back and forth, with change of direction and velocity (which any elastic, oscillating motion must involve) must be accelerated. This process requires force; and force acting through a distance is energy. The transmission or conversion of this energy is the function of so-called "acoustical" materials.

Finally, the scale (or dimension or proportion) of these phenomena must be remembered in any discussion of acoustics and acoustical materials. Materials must be considered on the molecular scale as well as the macroscopic scale of panels or constructions. The dimensions of amplitude, wavelength, sound pressure, acoustic energy, etc., must be considered in every problem. If this were done, the rather ridiculous misconceptions often associated with acoustical material design and use would be avoided. "Breaking up" of sound waves with textured paints or effective absorption of sound with thin flocked surfaces would not be considered seriously if the designer simply remembered the dimensions he is dealing with. A careful examination of the units of dimension or quantity will even cast some light on the hard-to-understand concepts of acoustics (note, for example, the discussion of impedance in Section V, A, 1).

D. UNITS AND DIMENSIONS

A simple analysis of the principal units and dimensions of acoustics may be helpful here. The human response to sound energy probably accounts for most of the units used in acoustics. Since sound is a vibration in an elastic medium, it involves energy and pressure. The pressure of an impinging wave front on the human eardrum may vary from an almost incredibly small threshold of less than 0.0002 dynes per square centimeter (dyn/cm^2) to more than 10^6 times that much (the so-called "threshold of pain"). And, as in other human sensations, there is not a simple, linear relationship between stimulus and response, but rather a logarithmic response; that is, the response varies as a ratio of the intensity of the stimulus. Acousticians have borrowed a unit from the electrical engineers—the Bel—to define the ratio of 10 to 1 between intensities or pressures. But this unit is too coarse for many masurements, so they have divided it into tenths and invented the "deci-Bel," abbreviated dB. The unit is rather anomalous, but its use has been so well established that change is probably impossible. Decibel means

$$10 \log_{10}(\text{value}_1/\text{value}_2)$$

It is used in acoustics to express levels above some arbitrary threshold. For example, sound pressure levels (SPL) are expressed as:

dB re 0.0002 dyn/cm² (or microbar)

Sound intensity levels (SIL) are expressed as

$$\text{dB re } 10^{-16} \text{ w/cm}^2$$

Sound intensity levels, then, vary as

$$10 \log_{10}(I_1/I_2)$$

Since intensity varies as the square of the pressure, sound pressure levels vary as

$$10 \log_{10}(P_1^2/P_2^2)$$

$$20 \log_{10}(P_1/P_2)$$

Table II indicates the relationship. More useful to the engineer, however, is Table III, which shows the levels of various familiar sounds.

To complicate the matter further, the ear responds in a complex manner to frequency. As Fig. 4 shows, we are somewhat more deaf to low-frequency sounds than to higher-frequency sounds. Only in the region of 1000 cps are the contour intervals fairly uniform. It is important to keep this in mind in analyzing the performance of acoustical materials. Perhaps this complex response has its advantages; it is very much more difficult to control low-frequency sound than higher-frequency sound, as later sections of this chapter will demonstrate.

TABLE II
RELATIONSHIP OF SOUND PRESSURE LEVEL TO INTENSITY

Relative intensity, units	Sound pressure level, dB
100,000,000,000,000	140
10,000,000,000,000	130
1,000,000,000,000	120
100,000,000,000	110
10,000,000,000	100
1,000,000,000	90
100,000,000	80
10,000,000	70
1,000,000	60
100,000	50
10,000	40
1,000	30
100	20
10	10
1	0

TABLE III
Typical Overall Sound Levels

AT A GIVEN DISTANCE FROM NOISE SOURCE   ENVIRONMENTAL

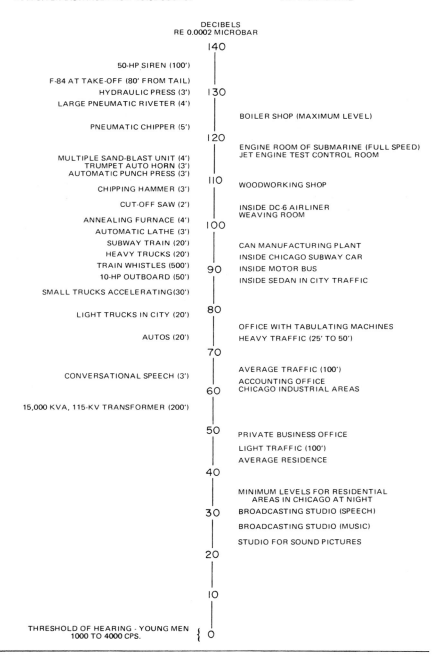

DECIBELS
RE 0.0002 MICROBAR

140

50-HP SIREN (100')

130
F-84 AT TAKE-OFF (80' FROM TAIL)
HYDRAULIC PRESS (3')
LARGE PNEUMATIC RIVETER (4')

BOILER SHOP (MAXIMUM LEVEL)

PNEUMATIC CHIPPER (5')

120

ENGINE ROOM OF SUBMARINE (FULL SPEED)
JET ENGINE TEST CONTROL ROOM

MULTIPLE SAND-BLAST UNIT (4')
TRUMPET AUTO HORN (3')
AUTOMATIC PUNCH PRESS (3')

110   WOODWORKING SHOP

CHIPPING HAMMER (3')

CUT-OFF SAW (2')   INSIDE DC-6 AIRLINER
                          WEAVING ROOM
ANNEALING FURNACE (4')
AUTOMATIC LATHE (3')   100

SUBWAY TRAIN (20')   CAN MANUFACTURING PLANT
HEAVY TRUCKS (20')   INSIDE CHICAGO SUBWAY CAR
TRAIN WHISTLES (500')   90   INSIDE MOTOR BUS
10-HP OUTBOARD (50')   INSIDE SEDAN IN CITY TRAFFIC
SMALL TRUCKS ACCELERATING (30')

80
LIGHT TRUCKS IN CITY (20')

OFFICE WITH TABULATING MACHINES
AUTOS (20')   HEAVY TRAFFIC (25' TO 50')

70

AVERAGE TRAFFIC (100')
CONVERSATIONAL SPEECH (3')   ACCOUNTING OFFICE
                          60   CHICAGO INDUSTRIAL AREAS

15,000 KVA, 115-KV TRANSFORMER (200')

50   PRIVATE BUSINESS OFFICE

LIGHT TRAFFIC (100')
AVERAGE RESIDENCE

40

MINIMUM LEVELS FOR RESIDENTIAL
AREAS IN CHICAGO AT NIGHT
30   BROADCASTING STUDIO (SPEECH)

BROADCASTING STUDIO (MUSIC)

STUDIO FOR SOUND PICTURES

20

10

THRESHOLD OF HEARING - YOUNG MEN   { 0
1000 TO 4000 CPS.

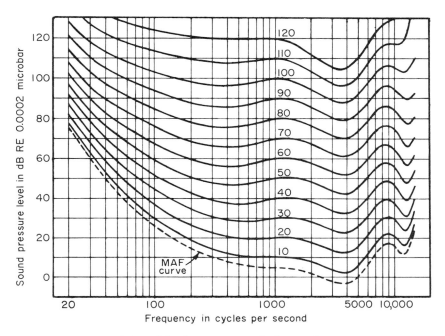

Fig. 4. Equal loudness contours for human hearing.

Probably borrowing from musical notation, the acoustician frequently uses the term "octave" in his work. However, he is usually interested in the octave only as a frequency ratio, not as a series of eight intervals. In this chapter, the octave is a frequency ratio of 2/1. Thus, the materials that transmit sound and vibration move twice as fast with each doubling of the frequency or each octave rise.

III. Absorptive Materials

A. Porous Absorptive Materials

The best-known acoustical materials are the absorptive materials. Porous, "fuzzy," fibrous materials, perforated boards, and similar building products are widely known as "acoustical." The normal furnishings in a room are also highly absorbent. For example, fabrics, carpets, cushions, and upholstery may be very effective absorbers.

1. Mechanism of Sound Absorption

The usual mechanism of sound absorption is a relatively simple process. The air contained within a porous matrix of fibers, granules, or particles of some sort, is "pumped" back and forth within a restricted space when sound energy reaches the medium. Whether the flow is turbulent or laminar, there are frictional losses as the air moves within the matrix. The frictional losses occur as heat, and the acoustic energy within the medium is reduced accordingly.

2. Absorption Coefficient

The amount of energy conversion is called the absorptivity or absorption of the material. It is usually expressed in coefficients that relate the energy converted to the total energy reaching the surface of the material.

If one square foot of an open window is assumed to transmit all and reflect none of the acoustical energy that reaches it, it is assumed to be 100% absorbent. This unit—one square foot of totally absorbent surface—is called a "sabin." Then the absorption of one square foot of an acoustical material is compared with this standard, and the performance is expressed in coefficients such as 65% or .65. As discussed in later sections, this rating method is not as simple or logical as it may appear, but workable measurement procedures have been devised to compare and rate absorbents and to predict their performance in actual use.

3. Structure of Porous Absorbents

The internal structure of most absorptive materials can be thought of as one or more of the three types shown in Fig. 5. The structure is rarely regular or uniform, and it may be a combination of all three types. Essentially, however, it serves to restrict the free flow of the contained air. The amount of energy conversion is determined by the resistance to air flow within the matrix (more correctly, to the "impedance," since the flow is ac rather than dc, and is rarely in phase with the pressure causing the flow; however, the principle remains the same).

The relationship between flow resistance and absorption is shown in Fig. 6. As might be expected, if the resistance is too low, frictional losses are low, and little energy conversion occurs. If resistance becomes too high, flow becomes so restricted and air motion so limited that frictional losses are again low. However, there is a relatively broad peak in

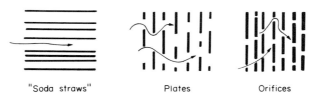

"Soda straws"   Plates   Orifices

FIG. 5. Internal structure of acoustical absorbents.

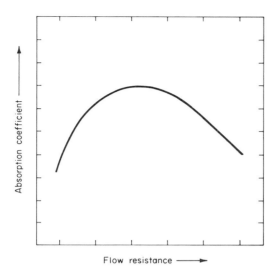

FIG. 6. Relationship between absorption and flow resistance of an absorbent material.

the flow resistance curve, and optimum flow resistance is not a sharply critical point in most materials.

It is important to remember that the maximum excursion of each molecule of air is usually relatively small; it is not necessary that the air enter or leave the absorptive medium, only that it move within the medium. Also, air movement can take place in any direction—vertically, horizontally, diagonally—within the matrix in which it is confined.

The porosity of absorptive media is usually related to the density of the media. However, this relationship is unique to each material or type of material. Density alone is not a meaningful indicator of the absorptive effectiveness of even a family of similar materials.

Fiber or particle size and orientation, too, affect the absorptive characteristics of materials. Multiple layers of ordinary fly screen, for example, would provide acoustical absorption, and the absorption would vary with the spacing between layers, orientation of strands, size of strands, etc. (5, 6).

Most absorptive materials have an extremely random fiber or particle orientation within the structure of the material. Air flow is usually through tortuous paths rather than simple tubes or orifices. Flow within most materials is probably quite turbulent. As might be expected, absorption within a material is related to the thickness of the material. Usually absorption increases with material thickness, but not in a simple, linear manner, as shown in Fig. 7.

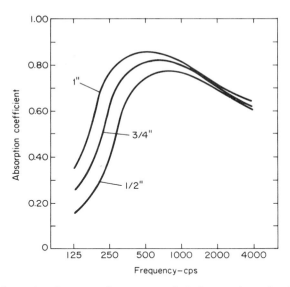

Fɪɢ. 7. Relationship between absorption and thickness of an absorbent material.

Equally predictable is a relationship between sound frequency and absorption, since motion of the individual molecules (reversals of direction of each molecule) is directly related to the frequency. Thus the dimensions of the material compared with the wavelength of the sound reaching it are significant.

Theoretically, maximum absorption occurs when absorber thickness is about one quarter of the wavelength of the lowest frequencies to be absorbed. Further, an optimum absorber has a graduated impedance or resistance, beginning with the impedance of air at the surface and increasing toward the interior of the material. In general, for optimum absorption, fairly fine strand or fiber sizes should be in a completely random ("haystack") orientation; or, if the structure is composed of granules or particles, their shape should be somewhat irregular so that the interstices between them vary in size, shape, and direction. The particles or fibers should not be large enough to reflect effectively even the highest frequencies of concern, but they must not be fine enough to make a matrix that is so dense that air flow is too restricted.

It is imperative that the internal structure be composed of interconnected pores and voids. Only open-cell structures are effective absorbers. Many plastic and elastomeric foams, and almost all glass and ceramic foams, tend to have closed, nonconnected voids. Air movement within them is very limited—almost nonexistent in most such materials— and they provide little or no sound absorption. A simple test of a material is to blow smoke through it. Those materials which pass no smoke will provide almost no absorption. If they pass smoke too easily, with little pressure, they may not have good absorption either.

From the standpoint of acoustical absorption alone, the material chosen must be of the proper porosity and thickness to absorb acoustical energy in the frequencies of interest. In practice, this usually means boards, panels, or blankets ranging in thickness from ½ to about 4 inches.

It should be obvious that mere rough textures, such as sand float finish on plaster or some of the so-called "acoustical paints" that repeatedly appear on the market to deceive the gullible, cannot be effective acoustical absorbers. The wavelength of even the very high frequencies of interest in most work is many times greater than the maximum dimension of such irregularities. But even more important, such products are not porous, and there is no opportunity for any air flow within them. Therefore there is no mechanism for sound absorption, and these products are almost totally useless. The engineer should be equally skeptical of the performance of very thin porous or textured materials. A thin, flocked surface over hard and impermeable materials is also almost useless except at very high frequencies. Very thin layers of porous foams have frequently been sold as absorbers. Some such materials are actually very good absorbers in thicknesses of one inch or more; but in the ⅛- to ¼-inch thicknesses in which they appeared on the market, they were quite poor absorbers in any frequencies below about 1000 cps.

4. Surfaces and Facings

In practice, the significant characteristics of acoustical absorbents include far more than just the acoustical absorption. Cost is the most significant parameter; strength, hardness, durability, cleanliness, weight, maintainability, fire resistance, moisture resistance, and appearance are among the other characteristics the engineer must look for. Hence, some of the best absorptive materials are not well suited for actual use unless modified in various ways.

One of the most effective modifications is to provide a surface of some sort and to depend upon the surface to protect the absorptive material and to receive the maintenance efforts, whether such efforts be painting, washing, or other normal maintenance procedures. Many acoustical absorbents, when used in exposed locations, are protected with porous or perforated facings of various types. The facings, then, introduce another significant factor. Any facing of any type will affect the acoustical performance of the material in some way. Usually the facings tend to degrade the high-frequency performance of the absorbent material (above 1000 cps), but they often improve the low-frequency performance. An explanation of this phenomenon may be gained from Fig. 8. This sound wave is diffracted as it strikes the perforated surface. When the solid or reflective areas between perforations become large enough, these areas reflect much of the sound which strikes them, particularly the higher-frequency energy. However, the lower-frequency sound, with its longer wavelength, can be diffracted around the reflective areas and provide the necessary pressure behind the surface to activate the air enclosed in the absorptive material behind the surface. In effect, a plate with a number of controlled orifices is inserted between the oncoming wave front and the absorptive material. The orifices offer minimal resistance to the lower frequencies but maximum resistance to the higher frequencies.

FIG. 8. Sound diffraction through a perforated plate.

The resistance of a perforated facing over an absorptive material must vary with the thickness and resistance of the absorptive material behind it to give maximum absorption over the widest range of frequencies (7). However, as the percent of open area decreases and hole spacing and size increase, low-frequency absorption tends to increase, and an increasingly sharp absorption peak appears.

Perforated facings used over commercial acoustical absorbents vary from about 5% to 40% open area, depending upon the thickness of the facing material, hole size, and hole spacing. Even very porous fabrics (such as speaker grille cloth) are used, and ordinary #16 fly screen is frequently employed for this purpose.

A common type of acoustical absorptive tile used in architectural work consists of a low-density fibrous insulation board with a thick, heavy, painted surface. Holes are drilled through the surface into the absorptive material below. Air flow takes place in all directions, but maximum air flow appears to be horizontal between the cylindrical voids in the structure.

Even a solid, unbroken film can be used over a good absorbent matrix if the film is thin enough, light enough, and flexible enough to impose little resistance between the impinging sound wave and the air in the matrix. Very successful acoustical tiles and panels are available with flexible films of Mylar, plasticized vinyls, etc., ranging in thickness from ½ mil to 2 mils. The films are not attached solidly to the entire surface they cover, since this would cause them to form a very rigid covering over the tiny openings in the surface of the material below. Rather, they are attached in the edges of the tiles or panels or in a few spots or in widely-spaced strips so that the entire film is free to flex and to be relatively limp. As mentioned earlier, the air at the surface of the panel need not enter the matrix, nor need the contained air flow out of the matrix. It is only necessary that the pressure of the oncoming wave be imposed on the contained air in the matrix to cause it to move. A thin, flexible, limp layer over the surface will permit this pressure transfer.

Since painting is a common, standard maintenance method, it is important that its effect on the absorption of the acoustical material be known. As explained above, a rigid, intimately-attached film will cause the material to reflect rather than absorb sound. For this reason, many acoustical tiles or panels are made with surfaces containing large openings—holes, fissures, etc.—to permit painting without degrading the sound absorption of the material. In practice, such surface openings should provide about 15% to 18% open area, and they should be large enough to prevent "bridging" or filling by normal paint used in normal application techniques.

Continuous films of controlled porosity over a confined air volume provide acoustical absorption, too. As air is "pumped" back and forth through the film, the acoustic energy converted to heat in this process can be appreciable. Occasionally, such films are made to adhere to a perforated sheet of metal, plastic, gypsum board, or other construction material to provide a type of absorber that functions somewhat as a resonator and a controlled resistance. Porosity of the film, dimensions and spacings of openings, and dimensions of the air space behind the film all affect the absorption of the structure (8). (See Section III, C, Resonators.) Absorption provided by such absorbers tends to be low when compared with most acoustical tiles, panels, or blankets, and the absorption curve (Fig. 9) tends to show the effects of the resonance of the air volume.

5. Mounting Methods

The method of attaching or supporting absorbents has an appreciable effect on their performance (9). Probably most materials are attached so

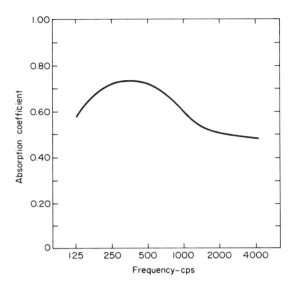

Fig. 9. Absorption of porous film applied to a perforated plate.

they adhere directly to a hard, impervious surface. In architectural practice, however, panels or tiles are often suspended on runners or furring strips, with an air space behind them. The effect of the air space is to increase the low-frequency absorption considerably and to degrade slightly the higher-frequency absorption. The absorption curve tends to rotate about the value at about 500 cps (Fig. 10). For this reason, published values of acoustical absorption of commercial materials are always related carefully to the method of application (8).

Depending upon the particular material being considered, the optimum air space behind the panel varies from about 2 to over 12 inches. However, little change takes place after the air space reaches 16 inches. As a result, the principal commercial testing laboratory for acoustical absorbents in the United States has standardized the air space at 16 inches for all tests on the No. 7 Mounting—the so-called mechanical suspension mounting system.

6. Types of Porous Absorbents

Hundreds of absorbent materials are in use today. For a complete up-to-date listing, refer to the current annual bulletin of the Acoustical Materials Association (8). The principal types of materials are summarized in Table IV.

In particularly difficult environments, the more conventional absorbents are usually unsuitable. Very high temperatures, corrosive fumes,

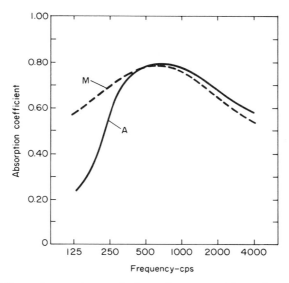

Fig. 10. Effect of mounting method on absorption: M, mechanical suspension; A, adhesive application to a reflective surface.

high humidity, dust, abrasion, impact and other physical abuse, and even very high sound pressure levels (> 150 dB) often eliminate the more common materials from consideration. Fortunately, however, materials or combinations of materials are available for use under such circumstances.

Metal "wools" made of stainless steel or copper fibers have been used

TABLE IV
COMMON POROUS ABSORBENTS

Material	Thickness, inches	Density, lb/ft³	Noise reduction coefficient[a]
Mineral or glass wool blankets	½–4	½–6	0.45–0.95
Molded or felted tiles, panels, and boards	½–1⅛	8–25	0.45–0.90
Plasters	⅜–¾	20–30	0.25–0.40
Sprayed-on fibers and binders	⅜–1⅛	15–30	0.25–0.75
Foamed, open-cell plastics, elastomers, etc.	½–2	1–3	0.35–0.90
Carpets	Varies with weave, texture, backing, pad, etc.		0.30–0.60
Draperies	Varies with weave, texture, weight, fullness, etc.		0.10–0.60

[a] Refer to Section III, E.

where high temperatures or corrosive gases are encountered. Unfortunately, they are very expensive and must be used in very thick layers to be effective.

Porous ceramic materials and some plastic-bonded mineral or glass wool fibers withstand humid conditions reasonably well. Foamed, open-cell urethanes are particularly useful in such environments.

Very few absorbents can withstand much abrasion or impact. A porous, sturdy facing is usually required to protect the absorbent in such applications.

Surprisingly, even many resilient materials with high temperature resistance do not perform well under sustained exposure to very high sound pressure levels. At about 160 dB (roughly, the level of the exhaust of a jet engine), binders burn out of most bonded mineral or glass wools; and the brittle, fragile structure of most boards or panels is subject to cracking or disintegration under the flexing and vibration that occur within the material at these levels. Open-cell (reticulated) urethane foams perform very well under these circumstances, and they are frequently used in such applications (where they are not exposed to hot exhaust gases or other high-temperature environments).

Massive, strong ceramic or porous masonry block materials are often used where sound pressure levels do not exceed 150 dB, but their absorption coefficients are relatively low when their structure is dense enough to resist the forces they are exposed to.

Porous, sintered metal is occasionally used for special applications, but it is extremely expensive, and it is rarely used in large quantities.

Damped metal panels or resonators are frequently used in these special applications, usually in conjunction with other absorbents (see Section III, B).

In special rooms, such as anechoic test chambers, tapered "wedges" (of various absorbent materials) up to 60 inches long are used. Their design is a highly specialized procedure (10) and should be undertaken with great care.

Occasionally, so-called "space absorbers" have been tried where the more usual panels, tiles, or blankets are not practicable (11). These free-hanging units may be thick panels (one type was 24 by 48 by 2 inches) of any of several types of absorptive material; hollow tetrahedrons formed of dense glass wool blankets covered with a thin vinyl film; two molded hollow wood fiber cones attached at their bases to enclose a large volume of air; hollow cylinders of glass wool with a perforated metal facing; or various other shapes. The units usually exhibit a high absorption per unit surface area, and they can be hung where required. However, they are relatively expensive, and they have not been used widely.

Most upholstery, seat cushions, fabrics, and clothing are absorptive. Published data are often available from manufacturers (particularly manufacturers of auditorium seats). People, the audience present in a room, are highly absorbent (12). Their absorption is always taken into account in design of critical spaces (13).

Rough estimates of the performance of absorbents can be calculated from known parameters of the materials; small-scale tests also yield valuable information. However, in design it is almost imperative that large-scale tests (preferably reverberation room tests) be conducted on large specimens of the material or construction being considered. Random incidence of sound, effects of mounting methods, edge effects, and other important variables are so significant that it is almost impossible to extrapolate from small-scale test results to performance in the field.

B. DIAPHRAGMATIC ABSORBERS

1. Mechanism of Sound Absorption

When a wave front impinges upon a panel, the panel vibrates at the same frequency as the sound reaching it. Since the panel material is never perfectly elastic, some energy is lost because of the inherent damping in the panel or the assembly in which the panel is used. This energy loss can be usefully employed as sound absorption in many cases.

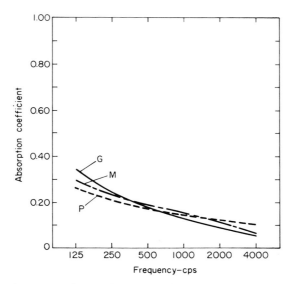

FIG. 11. Absorption of typical thin panel materials: G, window glass, double-strength; M, sheet metal, 26 gauge; P, plywood, ⅜-inch thick.

As might be expected, low-frequency sound will move panels more effectively than high-frequency sound, since the impedance match is usually much better between the air and the panel. High-frequency sound tends to be reflected without losing much of its energy to the panels normally used for most purposes (Fig. 11).

2. Types of Diaphragmatic Absorbers

In practice, thin sheets of metal, plywood, plastic, or even paper have been used as diaphragmatic absorbers. Typical units include vacuum-formed ceiling panels of thin styrene or vinyl, damped sheet metal, and even plywood. Because the absorption is significant principally in the lower frequencies, such absorbers are normally used only to supplement other absorption or to absorb specific low-frequency sound. This method has many applications, since it is often impractical to use the extremely thick layers of fibrous material required for good low-frequency absorption, but it is simple to use combinations of panels and blankets to provide good broad-band absorption.

One of the most ingenious applications today is a formed ceiling coffer consisting of light (26 gauge) sheet metal backed with a thick layer of mineral or glass wool. The metal surface acts as a light reflector, visible ceiling, and absorber (Fig. 12). It is possible to perforate portions of the metal panels to provide a low-impedance path for sound to reach the wool above and to give excellent broad-range absorption (Fig. 13).

Thin sheets of plywood over a confined air volume can also provide useful low-frequency absorption in building construction. If porous mineral or glass wool blankets are hung within the cavity behind the plywood surface, the absorption is appreciably increased. A design procedure for use of this type of construction has been worked out (14), and it can be quite useful for special applications such as theaters, music rooms, radio studios, and the like.

One remarkably efficient diaphragmatic unit commercially available today is a molded panel consisting of glass fiber, bonded with a plastic material to make a shell about ⅛ inch thick. The individual units are shaped into shallow, pyramidal vaults about 24 by 24 inches, and are erected on metal runners to provide a substantial air space behind them. They give a remarkably flat absorption curve, with unusually good coefficients from 125 to 4000 cps.

In general, practical requirements of strength, damage resistance, cost, and other characteristics of building and equipment materials tend to limit the applications of diaphragmatic absorbers. However, it is often wise to consider the inherent absorption of any such thin panels during design, both as a means of supplementing other absorption or as a major

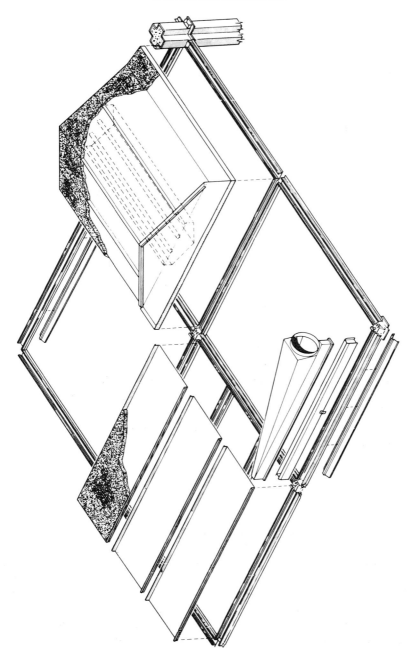

Fig. 12. An integrated ceiling system. (Courtesy of Inland Steel Products Co.)

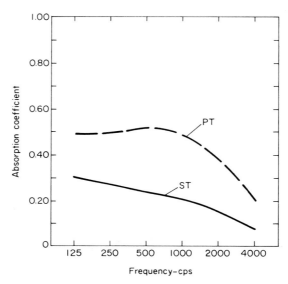

Fig. 13. Absorption of integrated ceiling system: PT, perforated top panels; ST, solid top panels.

consideration in spaces where low-frequency absorption may even be undesirable (music spaces, organ lofts, etc.).

Absorption varies with the mass and stiffness of the panels. Hence, it is difficult to calculate or forecast the absorption of panels unless all design details and dimensions of the completed construction are known accurately. In practice, prototypes are tested in a reverberation chamber to determine actual performance.

C. Resonators

1. Mechanism of Sound Absorption

Resonators (often called Helmholtz resonators) are cavities that confine a volume of air that communicates with the atmosphere by means of a small hole or channel in the surface of the cavity. If the dimensions of the cavity are very small compared with the wavelength of sound reaching the opening of the cavity, the resonator "tunes" to a specific frequency. The fundamental vibration of the confined air volume is a periodic air flow through the channel into and out of the cavity, and the air in the cavity acts as a spring. The kinetic energy of the vibration is essentially that of the air in the channel moving as an incompressible and frictionless fluid.

In practice, this type of absorber has limited application, since its peak absorption is a narrow band of the lower frequencies of interest in most sound control work. However, for applications where it is important to get high absorption of low frequencies, resonators can often usefully supplement other absorbents.

2. Types of Resonators

One of the most successful types of resonators used in the building industry is the ordinary concrete block with carefully designed slots cut into one face to form the channel which communicates with the hollow cells within the block. Most concrete masonry block used in building today is rather porous and somewhat absorbent. As a result, the blocks used as resonators have a distinct absorptive peak, usually in the frequencies between 100 and 300 cps, with some useful absorption in the frequencies above 300 cps. If an absorbent material such as mineral or glass wool is inserted in the cavities, the absorption peak is effectively broadened, and the absorption in the higher frequencies is increased significantly, as illustrated in Fig. 14.

Occasionally, resonators are attached to stacks, ducts, pipes, or other structures in which a strong, low-frequency tone must be attenuated. In such applications, it is often possible to obtain considerably more noise reduction with a resonator than with very thick layers of fibrous absorbents. The principle is similar to that of so-called side-branch resonators used for exhaust mufflers and other applications where nondissipative absorption is desirable. Design of such resonators is governed by the formulas for resonance of Helmholtz resonators, which can be found in almost any physics textbook. The frequency (or center frequency of the band of frequencies) to be attenuated is determined, and the volume and dimensions of the resonator and the orifice and neck or channel into the resonator are chosen to produce peak absorption at that frequency.

In most building applications, however, the dimensions of the cavity and the channel are governed by the dimensions of the wall structure into which they are built or by practical limits to the dimensions of the building units used. The design parameters that can be manipulated are normally the length and width of the slot or slots into the cavities.

It is usually imperative to choose practical units or materials with which to form the resonators, design proper openings or channels, and test the units in a reverberation chamber to compare their performance with the calculated performance. Modifying the measured performance tends to be an empirical process because of variables inherent in the materials, shapes, and other characteristics of the units.

FIG. 14a and b. See facing page for legend.

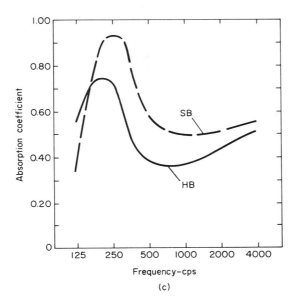

FIG. 14. Concrete block resonators; *a*, stuffed cells; *b*, hollow cells; (courtesy of The Proudfoot Co.); *c*, absorption of block resonators: SB, stuffed cells, HB, hollow cells.

D. Uses for Absorptive Materials

A discussion of the many aspects of sound control is available in the excellent books on this subject listed in the References. Only a very brief review of the principal uses for sound absorbents, and of their effects, is given here.

It is very important to recognize that most absorbents are very *poor* sound barriers. They are usually lightweight and porous—quite the opposite of good barriers (see Section IV). It is almost useless to apply thin, porous tiles, blankets, or panels over walls, for example, to minimize noise transfer through the walls. Absorbents should be thought of as "blotters;" like blotters, they "soak up" sound, but they do little to reduce its passage through them.

1. Functions of Absorbents

Acoustical absorbents are used

    a. To reduce sound pressure levels in rooms, enclosures, and the like
    b. To prevent reflections, particularly in acoustically critical spaces such as auditoriums and music halls

c. To control reverberations in any enclosure

d. To provide a controlled acoustic resistance; e.g., in architectural work, as wind-screens over microphones.

2. Location of Absorption

In general, absorbents should be located

a. As near the offending sound source as possible

b. On the surface producing unwanted reflections

c. On surfaces not required for helpful reflections

d. In locations where they control or restrict air flow as required.

In practice, this usually means that they are found on ceilings, walls, and floors of rooms; on panels surrounding noisy equipment; within the cavities between wall or partition surfaces, and the like.

Most discussions of the effect of absorption assume a diffuse, random sound field and random incidence of sound on all surfaces. In practice, this condition rarely exists. In an enclosure, for example, strong standing waves may exist between parallel surfaces, and absorption on other surfaces may have much less effect than the assumption of a random, diffuse field would predict. Thus, the effective absorption contributed by absorbents is usually lower than that calculated. To obtain maximum effect of a given area of absorption, the material should be distributed widely and in a somewhat random manner.

3. Design Formulas

It is possible to calculate with reasonable accuracy the effect of introducing absorption into a space. Three formulas are of particular interest to the engineer.

a. Noise Reduction

$$NR = 10 \log_{10}[(A_o + A_a)/A_o] \tag{2}$$

where A_o is the original absorption present in sabins, A_a is the added absorption in sabins, and NR is the sound pressure level reduction in decibels.

Note: The surface area in square feet multiplied by the absorption coefficient equals the sabins of absorption. When added absorption covers an existing surface, the coefficient of the added absorption must be reduced by the coefficient of the existing surface covered by the added absorption.

It is obvious that the effect of the added absorption in a highly absorptive space is smaller than its effect in a nonabsorptive space. In practice, less than 10 dB noise reduction can be accomplished in most installations by the introduction of absorption alone.

b. Reverberation Time

$$T = 0.05(V/A) \tag{3}$$

where T is the time in seconds for the sound pressure level to decay 60 dB after the source ceases, V is the room volume in cubic feet, and A is the total absorption in sabins within the space.

c. Standing Wave Ratio

In an enclosed space such as an anechoic test room, some energy is reflected from the enclosing surfaces, however absorbent they may be. A so-called "standing wave," or nodes and antinodes of sound pressure, will be found within the space.

If it is important that sound pressure levels throughout the room do not vary by more than a given amount, it is necessary to know the absorption coefficient required for the surfaces of the enclosure.

By definition

$$\text{SPL}_{max} - \text{SPL}_{min} \text{ in dB} = 20 \log_{10}(\text{Pressure}_{max}/\text{Pressure}_{min}) = n \tag{4}$$

Then the reflection coefficient of the surfaces,

$$r = (n - 1)/(n + 1) \tag{5}$$

Then the absorption coefficient (ratio of energy absorbed to the incident energy),

$$\alpha = 1 - r^2$$

or $\tag{6}$

$$\alpha = 4/[n + (1/n) + 2]$$

when α is determined by impedance tube measurements.

In large rooms, the absorption coefficient is somewhat larger than α, varying from almost twice as large at small values of α to about 1.25 as large at high values of α. This is to be expected, since the impedance tube measures only normal incidence energy effects rather than random incidence as in a room.

For very high absorption (small standing wave ratios), very thick absorption or tapered absorptive "wedges" are required for good low-frequency absorption. The length of such wedges must be approximately one-quarter wavelength (10).

E. Tests and Test Methods

Performance data on materials can be obtained from small-scale tests on a few square inches of material and from full-size mock-ups of assemblies. The more important test procedures follow.

1. Small-Scale Tests

a. Flow Resistance Test

One simple test method of forecasting the absorption of a porous absorbent is the flow-resistance test (ASTM Test Method No. C 522-63T). The flow resistance in itself is almost meaningless, but in a given family of materials it may indicate whether the structure of the material is roughly correct to provide good sound absorption. (Refer to Section III, A, 3.)

The test is often used as a quick quality-control test for mass-produced materials. By "calibrating" flow resistance for a type of material against results of acoustical absorption tests on the same material, a fair indication of performance can be obtained.

b. Impedance Tube Test

The impedance tube (ASTM Test Method No. C 384-58) is a sophisticated version of the classical Kundt's tube encountered by every high school physics student. A sound source with controlled frequencies and steady, controllable pressure levels closes one end of the tube; a test specimen covers the inside surface of a solid, reflective termination at the opposite end. A small probe microphone can be moved within the tube to measure maxima and minima. Then absorption can be calculated. (Refer to Section III, D, 3.)

The method is limited to the higher frequencies, and two or more tubes of different diameters are frequently used to cover a wider frequency range. Unfortunately, the absorption measured is for normal incidence only, and values must be corrected to provide random incidence coefficients.

Since about 50% of the sound impinging upon a surface in a completely random sound field arrives at angles of more than 58° from normal incidence (Fig. 15), it is highly likely that random incidence absorption varies greatly from normal incidence absorption. Further, the significant effect of a mounting method on absorption cannot be measured in the impedance tube. Nevertheless, the method is simple, accurate, and repeatable; and results can be calibrated against reverberation-room results on the same material. The impedance tube is a good research tool.

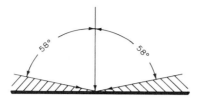

FIG. 15. Effect of sound incidence. One half of random incident energy arrives at angles greater than 58° from normal incidence.

2. Large-Scale Tests

The reverberation-room method (ASTM Test Method No. C 423-66) is the principal large-scale test. For most applications of absorbents, the information required is the random incidence absorption when the material is applied by the mounting method to be used in practice.

The coefficients resulting from the test are often called the "Sabine coefficients," since they are computed from the effect of the test specimen on the sound decay in the room, using the Sabine reverberation-time formula. The results of this test method (rather than small-scale test methods) probably relate more accurately to the performance of the materials in actual job installations. The test is used principally to obtain values used by architects, engineers, and designers in the building and product development fields.

Sound absorption coefficients are usually determined at 125, 250, 500, 1000, 2000, and 4000 cps. A simple, single-number value often used in practice is the "noise reduction coefficient." This is the arithmetic average of the absorption coefficients at 250, 500, 1000, and 2000 cps. It has no physical meaning and should not be used indiscriminately. However, it is a reasonably useful means of comparing *similar* materials and of predicting the effect of the material in reducing general, broad-band noise within ordinary rooms, offices, and the like.

IV. Sound Barriers

The most important acoustical materials are those which reflect, contain, or "isolate" sound. Although the engineer may deliberately use a panel as a diaphragmatic absorber (see Section III, B) or as a reflector, he is usually interested in it as a barrier to contain acoustic energy or to block transmission of air-borne sound from one space to another. Containing sound, providing a barrier against its transmission through air, is undoubtedly the major problem in most sound control work. The walls of rooms, gear housings, airplane cabins, and auto bodies are examples of sound barriers.

For a more complete treatment of the subject of the use of sound barriers, refer to the references and bibliography. This section is intended only to demonstrate that all materials are, indeed, "acoustical" and that nonabsorbent materials are as important to the engineer and designer as the absorbent materials.

It is well known that a wall or heavy enclosure will serve as a very effective barrier against air-borne sound transmission. Although any surface reflects some of the sound that reaches it, only heavy, air-tight surfaces are significantly effective in containing or "stopping" sound. The more massive and airtight the surface, the more effective sound barrier it is.

It is very important to recognize that most acoustical absorbents are very poor sound barriers. They are porous and light weight—quite the reverse of what is required to reflect or isolate sound. Therefore, it is usually quite useless to apply acoustical tile, for example, over a wall to reduce sound transmission through the wall. The added mass is trivial, and the innumerable "holes" through the absorbent provide good paths for sound transmission. Only the energy that is absorbed is eliminated, and this will usually be less than 5 dB. When, for example, isolation of 30 to 60 dB is required, absorbents are not very useful.

A. MECHANISM OF SOUND ISOLATION

When a wave front reaches a barrier, the barrier is set into motion. The barrier, then, becomes a sound source, and sets into motion the air on the other side. Some of the energy is transmitted to the air on the opposite side of the barrier; some of the energy is reflected back toward the source; and some is lost in moving the partition (Fig. 16).

Even a very thick and massive barrier moves slightly. However, as we know instinctively, the motion is small, and considerable force is

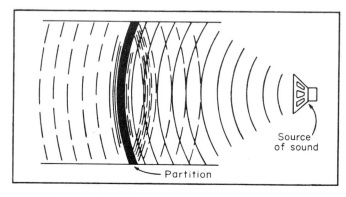

Source of sound

Partition

FIG. 16. Effect of incident sound on a solid barrier.

needed to move a heavy barrier. But even a tiny hole offers a path of low resistance through which sound transmission occurs readily.

Since the barrier moves with an oscillating and accelerated motion, force is obviously required to initiate and sustain the motion. The partition has mass; it is accelerated by the force or pressure of the impinging sound. wave front. Therefore, it is possible to analyze its motion mathematically.

From the laws of motion, we know that force equals mass times acceleration. The instantaneous kinetic energy of the moving partition is proportional to $\frac{1}{2}MV^2$, where M is the mass, and V is the velocity. Therefore, we know that more force (pressure) and more energy are required to vibrate a panel at higher frequencies than at lower frequencies. With each octave increase in frequency, the sound energy increases four times (in proportion to the square of the velocity) for a given panel mass.

For each doubling of the mass of the partition, the pressure (force) must increase by two times to maintain the same motion; and, since energy is proportional to the square of the pressure, the energy increases by four times.

Thus, the energy expenditure to maintain a given amplitude of motion for a panel increases by 6 dB per octave frequency increase and 6 dB per doubling of the mass per unit area of the panel. (Note: Relative

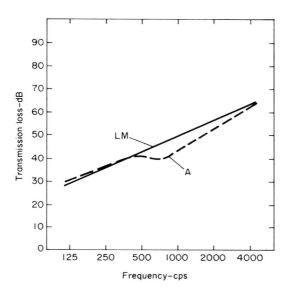

FIG. 17. Effect of coincidence on transmission loss of a stiff panel weighing 10 lb/ft²: LM, single layer, lamp-mass panel; A, actual stiff single-layer panel.

levels in dB are expressed as $10 \log_{10}$ of the ratio of values. Thus, $10 \log_{10}$ $4 =$ approximately 6 dB.)

B. SOUND TRANSMISSION LOSS

The ratio of the sound energy incident upon one surface of a partition to the energy radiated from the opposite surface is called the "sound transmission loss" (STL) of the partition. The actual energy "loss" is partially energy reflected (back toward the source) and partially heat (internal heat losses within the partition). Figure 17 shows the relationship between the transmission loss through a solid panel and the frequency.

1. Limp Mass Law

On the assumption that the partition is a "limp mass," moving essentially as a piston, the sound transmission loss for randomly incident energy, excluding losses at the edge of the panel, is calculated as

$$STL = 20 \log_{10} W + 20 \log_{10} F - 33 \qquad (7)$$

where W is weight in lb/ft², and F is frequency in cps.

The limp mass is actually a naive assumption rarely valid in a panel in actual use. The panel moves in a more complex manner, depending upon its stiffness. Often a significant "shear wave" (comparable to the transverse waves created in a vibrating string) occurs in the panel (4). When the velocity of this shear wave (determined by the stiffness of the panel) coincides with the component of velocity of the incident sound wave in the air, sound transmission loss through the panel is sharply reduced. Theoretically, it drops to zero, but internal losses (damping) within the panel provide appreciable attenuation. Actually, the coincidence effect results in a "plateau" in the STL curve, quite different from the simple 6-dB-per-octave limp mass curve. See Fig. 17.

Thus, the performance of a panel of a given material varies not only with the surface mass but with the elasticity or stiffness of the panel. Light-weight, stiff panels tend to behave much more poorly than their mass alone would indicate. Dense, limp materials, such as soft lead, behave nearly according to the mass law throughout much of the frequency range.

2. Double-Wall Construction

It is apparent that two serious natural limitations exist in all real materials that might be considered for sound barriers:

a. Truly limp materials have little use in the construction of most enclosures.

b. If a doubling of mass produces only about 6 dB improvement in sound transmission loss, a panel becomes prohibitively heavy when very high transmission loss is required.

Fortunately, a given mass of material may be used in a way that appreciably improves its sound transmission loss throughout most of the significant frequency range. If the mass is divided into separate layers with no rigid connections between layers, a substantial increase in performance occurs (Fig. 18).

The layer of air between layers of surface material (unless the air layer is very thin) is limp enough to provide poor energy transfer from one surface to the other. The shear wave in one surface is only very inefficiently coupled to the opposite surface; and the pistonlike motion of one surface is cushioned by the soft layer of air between layers of material. The impedance match between the surfaces and the entrapped air is very poor except at certain resonant frequencies; and even at those frequencies, internal damping provides substantial attenuation.

If an absorbent blanket is placed in the air space between layers, the air must "pump" back and forth through the absorbent material, further attenuating the energy.

Theoretically, each doubling of the air space between surface skins

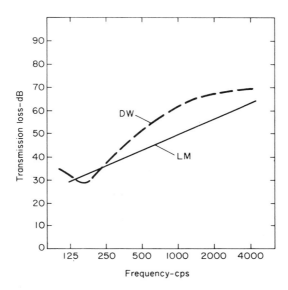

FIG. 18. Transmission loss for double-wall panel: DW, double-wall, consisting of two separate skins, each weighing 5 lb/ft²; LM, single-layer, limp-mass panel weighing 10 lb/ft².

of a double-wall partition should improve sound transmission loss by about 6 dB. Actually, the gain is somewhat less than this in actual constructions—more like 5 dB or less.

Where structural requirements preclude total separation between surface layers, a relatively soft inner layer may be used. Its effectiveness depends upon its shear modulus. Very low shear modulus materials permit each panel to move in shear, somewhat independent of the opposite panel. As a result, the panel tends to approach its limp mass sound transmission loss potential (4).

The necessity for the barrier to be airtight cannot be overemphasized. Even a tiny hole through a barrier constitutes a major leak. Obviously, the mass of the air in a hole is trivial, and the hole acts as a very low resistance pressure release and energy transmission path. A hole of one square *inch* area will transmit almost as much energy as 100 square *feet* of a wall rated at 40 dB transmission loss. Thus, open shrinkage cracks, the perimeters of doors, and similar apparently insignificant openings common to building construction can possibly vitiate the effect of an entire wall as a sound barrier.

C. Tests and Test Methods

At present, there is no method of determining sound energy flux directly. Rather, sound pressure level measurements are made, and energy is inferred from such measurements.

Sound transmission loss (ASTM Test Method E 90-66T) is determined by building a partition or barrier between two spaces, with a sound source in one space. The sound pressure level is measured in each space and the noise reduction is determined (SPL in source room minus SPL in receiving room). The absorption present in the receiving room is determined by measuring sound decay and calculating the absorption by the Sabine formula.

Then the sound transmission loss is calculated by the formula

$$\text{STL} = \text{NR} + 10 \log_{10} S - 10 \log_{10} A_R \qquad (8)$$

where NR is the noise reduction between source and receiving room in dB, S is the area of sound-transmitting surface of the test specimen in square feet, and A_R is the total absorption of the receiving room in sabins.

Because architects and engineers have a strong predilection for single-number ratings for materials and constructions, a scheme has been devised for comparing the actual STL curve with a standard contour (related to the importance of isolation required at various frequencies). This contour is roughly the inverse of the ear's sensitivity to levels at various frequencies—see Section II, D. The STL value where the contour intersects 500 cps is called the Sound Transmission Class of the construction [see ASTM E 90-66T, "Determination of Sound Transmission Class (RM 14-2)"]. It is important to remember, however, that only the complete STL curve provides sufficient information to characterize adequately the performance of an acoustical barrier.

It is possible to forecast roughly the sound transmission loss to be expected from a panel, knowing its mass alone. However, as discussed previously, the coincidence dip, structural separation, and construction details affect performance significantly. Most manufacturers of materials and constructions, and various agencies and trade associations publish current STL data. Most of the sources listed in the references and bibliography also contain considerable information and data on various materials.

D. PERFORMANCE OF TYPICAL BARRIERS

The range and magnitude of sound transmission loss performance for various materials may be inferred from Table V for typical panels used in building and product design.

TABLE V

SOUND TRANSMISSION LOSS FOR COMMON BUILDING MATERIALS

Material	STC Rating[a]
$\frac{1}{4}$-inch steel plate	36
$\frac{1}{4}$-inch plate glass	26
$\frac{3}{4}$-inch plywood	28
4-inch brick wall	41
6-inch concrete block wall	42
$\frac{1}{2}$-inch gypsum board on both sides of 2- by 4-inch studs	33
12-inch reinforced concrete wall	56
14-inch cavity wall:	
8-inch brick, 2-inch airspace, 4-inch brick	65

[a] Refer to Section IV, C.

V. Isolation and Damping Materials

Since all materials are somewhat elastic, all materials transmit sound and vibration to some degree. Many materials used by engineers and designers are highly elastic; hence, such materials may transmit sound and vibration readily. Often this characteristic of materials is a serious nuisance. Minimizing such transmission often presents a greater challenge than any other aspect of noise control.

A. Mechanism of Isolation and Damping

An obvious approach would be to interrupt the transmission path, and this is often a practical and successful approach. However, methods of accomplishing this appear not to be so obvious. Many engineers and designers appear to misunderstand the mechanism of vibration isolation.

Generally, there are two problems in vibration control:

a. Prevention of energy transmission between the source and the surfaces that radiate the sound and vibration.

b. Dissipation or attenuation of the energy somewhere in the structure.

1. Impedance Mismatch

The solution to the first problem requires that the structure be assembled in such a way that energy transfer through connections is inefficient. The term often used to describe this process is "impedance mismatch." Unfortunately, the term "impedance" seems to convey little information to many engineers. An examination of this term may explain how this concept can be used in design.

The acoustical impedance of a material is found by multiplying its density by the velocity of sound in the medium; this gives an impedance unit such as lb/in.^{2}-sec or in.-lb/in.^{3}-sec. The first expression represents force per unit area per second; the second represents energy per unit volume per second. Both indicate a rate at which force can be applied per unit area or energy can be transferred per unit volume of material. In other words, some materials cannot accept energy as fast as others. A good, homely analogy would be the effect of trying to hammer air or stone; it is simply impossible to hit the air as hard as the stone with a hammer. It is pretty difficult to make much noise hammering the air, but it is easy to make at lot of racket pounding a stone.

2. Resilient Mounting Methods

In vibration control, low-impedance materials are inserted between high impedance materials to interrupt the transmission paths. Even if the interrupting materials are highly elastic, they cannot transfer the energy of oscillating or vibrating source fast enough to transmit much noise or vibration. Thus, a steel spring can support a vibrating machine on a concrete base without transmitting much energy between them. The spring simply stores most of the energy which it accepts, transmits a little, dissipates a little, and returns most to the vibrating system with each cycle.

If a mass, supported on a perfectly elastic spring resting on an infinitely stiff and massive support, were set into vibration, it would oscillate at a rate determined only by gravity and the spring rate (the load required to produce unit deflection) of the spring. The frequency of this oscillation ("natural frequency") would be

$$f = 3.13(1/d)^{1/2} \tag{9}$$

where f is frequency in cps and d is static deflection of the spring in inches under the load imposed (determined by the stiffness of the spring).

If the spring were truly perfectly elastic, the mass would vibrate indefinitely at the natural frequency of the system (if the base were truly infinitely stiff and massive). The spring would simply store the kinetic energy of the system as potential energy and return it to the system with each cycle. However, no components of this nature exist. Internal damping losses within the system would eventually cause the free vibrations to cease unless outside force was applied to the mass. The decay of the vibration will, of course, be roughly logarithmic, since each oscillation will be damped by some ratio of the energy or amplitude of the previous oscillation (Fig. 19).

To accomplish perfect storage and return, the spring or elastic material would have to be perfectly elastic. Obviously, such a material does not exist. Some energy is lost in each cycle in internal dissipation within the spring. This loss is called "damping." (Of course, some energy is transmitted to the support; this, too, is ultimately dissipated and lost to the system.)

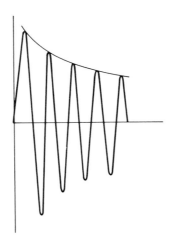

Fig. 19. Decay of a damped oscillation.

3. Damping

Damping is a means of dissipating or attenuating vibrational energy. In a sense, the damping mechanism is the opposite of resilient isolation. Instead of storing energy and returning it to the vibrating system, damping materials accept the energy and convert it to heat. The two mechanisms should not be confused.

Most vibrating systems can be thought of as a mass supported on a spring with an attached dashpot or other damping device (Fig. 20). The damping mechanism may be any of several types:

a. *Viscous.* The damping force is directly proportional to the velocity of a fluid, as, for example, in a dashpot using fluid forced through an orifice.

b. *Hysteresis.* The damping is "structural," depending upon displacement and having little frequency effect.

c. *Coulomb.* The damping is due to friction.

Nearly every text on dynamics contains a detailed treatment of vibration and damping. Vibration isolation and control are covered at length in the references following this chapter. However, the engineer and designer need know only a few basic concepts to obtain a reasonable understanding of damping and damping materials.

For several reasons, damping is a useful characteristic of all materials. The two most important reasons are as follows:

a. If force is continually applied to the resiliently supported mass, the system will continue to vibrate at a given rate and amplitude. Should the system be vibrating at its natural frequency, it will resonate—that is, the amplitude of vibration will increase with each cycle and the energy transmission to the base will increase constantly. In a perfectly elastic system, this increase would approach infinity.

Figure 21 shows how the transmissibility of vibration varies with the ratio of the forcing or driving frequency of the resiliently supported mass to the natural or resonant frequency of the resiliently mounted system. At a ratio of 1.0, resonance occurs. No isolation occurs for ratios less than 1.4.

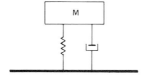

Fig. 20. A simple vibrating system with damping: M, mass.

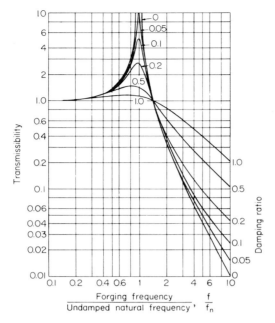

FIG. 21. Transmissibility of a vibrating system.

Damping reduces the magnitude of the resonance, but it also reduces the isolation provided by the system.

Actually, very dangerous resonance conditions can and do occur in practice. Fortunately, the inherent damping in any system will eventually limit the motion of the system to some maximum. For resiliently supported machines, there is usually a brief period of resonance as rotational speed increases from zero to operational rates. Either positive restraints must be provided or internal damping of the system must limit the motion. Of course, the greater the damping, the less the isolation. Too much damping degrades the performance of the mounts, whereas too little may permit undesirable movement in the system.

So-called "shock absorbers" on an automobile are typical of practical dampers. They limit motion of the sprung masses, and they restore the system to stability.

b. If a panel of elastic material—metal, wood, glass, etc.—is set into vibration by almost any source, it will radiate sound. If such a panel is part of the housing for a rotating or vibrating machine, for example, the panel can act as an efficient loud-speaker, amplifying the sound of the machine many times. Or a panel, if struck, will "ring" for some time. Fortunately, these vibrations can be restricted and damped by the proper design of the panel and by the application of damping materials that dissipate the energy as heat.

B. Types of Resilient Mounting Materials

Cork, hair-felt, bonded glass fiber boards, solid elastomers, and foamed plastics and elastomers are often used for resilient mounting systems. Usually, they are used in compression, either as pads or blocks of various sizes and thicknesses. The elastomers are often used in shear, since their performance is usually more truly elastic in this mode.

Internal damping is very high in many of these materials, particularly the granular and fibrous products. As a result, rarely can they provide much vibration isolation for frequencies below about 10–12 cps.

Most such materials exhibit strongly nonlinear stress-strain characteristics. Usually they are "hardening" springs, with the spring rate increasing with the deflection. Occasionally, they are "softening" (or even collapsing) springs, with spring rate decreasing or reaching a constant value. The dynamic performance of such materials is usually far different from their static performance; usually the spring rate is much higher under dynamic loading. A typical curve is shown below.

The large area within the "hysteresis loop" represents the energy loss per cycle; this, of course, is the damping provided.

Most such materials may be loaded from 1 lb/in.² to over 50 lb/in.². If loaded too heavily, they tend to fatigue and to experience permanent

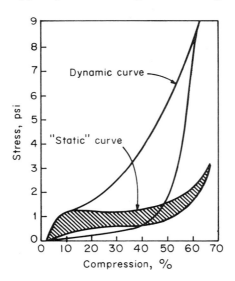

FIG. 22. Dynamic response vs. static response of a typical high-damping elastomer.

"set" or deformation. The useful life for these products appears to be about 20 years, although cork and some natural rubbers tend to harden and lose resiliency in much less time.

Loadings on these materials must be reduced, often to one-half, when they are subjected to shock.

For very high unit loadings and continuous shock loads, special materials are available. For example, multiple layers of fabric impregnated with Neoprene are laminated into sheets or pads. Such products may be loaded to 500 lb/in.^{2} or more.

C. Types of Damping Materials

Materials used solely to damp vibrations are usually viscoelastic substances, ranging from filled bitumens to specially formulated elastomers. Some of the new polymer plastics are particularly effective.

Materials are available in sheet form for adhesive application or for vulcanizing to metal; as liquids or thick fluids for spray or trowel application; and as tapes, often with contact adhesive already applied.

Flexing of the damping materials, either in tension and compression or in shear, provides the mechanism for energy dissipation. Intimate contact between the material and the structure to be damped is imperative.

Generally, the materials must be applied in a layer approximately equal in thickness or weight to that of the panel being damped if appreciable damping is required. Usually the optimum location for damping is at areas of maximum movement.

Most efficient use of the damping material occurs if it is used as a constrained layer; that is, the material is sandwiched between the panel to be damped and a relatively rigid layer such as thin sheet metal above. This forces the damping material into shear as the panel vibrates, dissipating substantially more energy than when the material acts simply in extension and compression.

Formulation of damping materials is a relatively complex process. Good and poor materials may be very similar in composition. Often the plastic or elastomer is loaded with heavy, inert, granular materials such as lead powder or galena to increase weight, or with mica to introduce Coulomb damping by means of friction between the particles suspended in the binder. However, simply adding such materials randomly may give very disappointing results. Usually it is best that the fillers be of such particle size and type that they actually rub together within the binder matrix.

A fair amount of damping can be accomplished by filling the space between surfaces or faces of a wall with a heavy, granular material, such as sand. When poured into the voids of a masonry block wall, for example, sand provides appreciable damping with a noticeable increase in sound transmission loss. However, sand has limited usefulness in most damping applications, and it is often a nuisance to use.

Carpets and carpet pads, cushions, and various "lossy" blanket materials are also effective shock absorbers and panel dampers. Some acoustical absorbent materials, if intimately adhered to panels, are reasonably good, too.

D. TESTS AND MEASUREMENTS OF DAMPING EFFECTIVENESS

There are many means of measuring and expressing damping effectiveness. Some of the more common criteria follow.

1. Decay Rate in dB per Second

This criterion is perhaps the most useful in evaluating subjective improvement. The effect of damping is evident in that the "ring" of panels is eliminated, radiation from vibrating surfaces is reduced, and other noticeable effects are mitigated (as when reverberation in a room is reduced).

The "Geiger Thick-Plate Test" (16) procedure is a simple means of measuring decay rate. The method lends itself to quick comparisons between materials and is well suited for research and development work and quality control.

2. Percent of Critical Damping

Critical damping is the damping necessary to just prevent oscillation. The damping ratio, C/C_c, is one means of expressing this criterion. Here C_c is the critical damping coefficient, and C is the damping coefficient of the material under consideration. The values themselves are not very informative unless compared with or calibrated against some other criteria. However, they are very useful in comparing the performance of products or in predicting performance of damped structures.

3. Loss Factor η

Damping is also often expressed (17) as the loss factor, η:

$$\eta = (1/2\pi)(D_0/W_0) \tag{10}$$

where D_0 is the total energy dissipated in the system as the result of damping during one cycle, and W_0 is the total energy of vibration of the system.

The value η is related to the damping ratio as follows:

$$C/C_c = \tfrac{1}{2}\eta \tag{11}$$

and η is also related to the resonance sharpness, Q, of the system as follows:

$$Q = 1/\eta \tag{12}$$

where

$$Q = f_0/\Delta f_0 \tag{13}$$

Here f_0 is the frequency of resonance, and Δf_0 is the number of cycles between the two frequencies at either side of resonance, where the intensity level is 3 dB down. The value η is related to the decay rate of a free vibration as follows:

$$\Delta t = 27.3\eta f \tag{14}$$

where Δt is the decay rate in dB/sec, and f is the frequency in cps.

There are many ways of measuring these various parameters. It is important only that the equipment used for dynamic measurements be sensitive enough to measure accurately without appreciably affecting the motion of the system. Various test arrangements are described in the references and bibliography for this section.

Some materials depend upon combinations of viscous, Coulomb, and hysteresis damping. For most applications this is desirable, but for some applications possibly only one of the damping mechanisms is desirable.

It is very important that tests be conducted on the materials to determine their damping characteristics. Often apparently very similar materials vary widely in performance, particularly with changes in temperature and frequency.

E. Performance Characteristics of Damping and Resilient Materials

Useful damping materials have decay rates from as low as 5 dB/sec to over 80 dB/sec, and from ½ to 20% of critical damping. Damping ratios range from a fraction of 1% to over 20% of critical damping.

Acoustical impedances for various typical materials are shown in Table VI.

VI. Conclusions

Modern society and modern technology are forcing upon us a growing awareness of our acoustic environment. New and catchy phrases such as "noise pollution" are coming into use. Today it is even possible to obtain information on the acoustical output or performance of machines, equipment, and materials from almost any established manufacturer; in the past, such requests were often greeted with blank stares.

The rising sound output from ever more powerful machines, the

TABLE VI
ACOUSTICAL IMPEDANCE OF VARIOUS MATERIALS

Material	Acoustical impedance, lb/in.²-sec
Cork	165
Pine	1,900
Water	2,000
Concrete	14,000
Glass	20,000
Lead	20,500
Cast iron	39,000
Copper	45,000
Steel	58,500

horrendous "sonic boom" produced by supersonic planes, and the increasing level of sound emanating from more active and crowded population centers are forcing us to "do something about" noise.

Noise is simply unwanted sound. The problem, then, is to control or eliminate it. Sound originates with a source, travels via a path, and reaches a receiver. However, for the most part, the engineer or scientist in research, development, design, or construction can only work on the path—the materials or media by which the sound or vibration is transmitted.

The nostalgia that leads us to long for "the good old days" and causes us to complain that "they don't make them like they used to" is not very helpful. The good old days didn't have the fantastic amount of powerful equipment common to almost every building today. Neither did they have 50-story apartments, 400-hp autos, or jet planes. These things are now a part of everyday life; we must learn not only to live with them but to make them acceptable in our lives.

Engineers tend to be single-minded people. They think in terms of greater horsepower, higher light levels, taller buildings, faster planes, and greater production. Unfortunately, they frequently overlook the increase in undesirable "side effects" which accompany the technological "improvements."

Noise levels have been rising for a long time. A generation ago the standard solution to most acoustical problems was to apply something somewhere, somewhat like putting a bandage on a small wound. Today such an approach is hardly acceptable; the small wound has become a massive hemorrhage.

It is imperative that the engineer remember that every material he uses is an acoustical material, that it will affect the acoustical output or environment of the product or space with which he is working. The acoustics of the problem cannot be left until tests of the prototype indicate a serious problem. By remembering at the outset that the acoustical characteristics of any material are related to the fundamental physical characteristics of the material, the engineer can control the acoustics of the project from the very beginning. Perhaps a molded plastic part, rather than a stamped sheet-metal cover would be better; possibly a different configuration would avoid the problem. Possibly the various functions of the whole device can be integrated into a single, multipurpose solution which exploits all of the characteristics of the materials used in the construction of the device. The ingenious modular ceiling system shown in Fig. 12 (Section III, B, 2) is a good example of this approach. The glass wool required for thermal insulation serves also as a fireproofing blanket, as sound absorption, and as vibration damping. Even the sheet-metal panels were chosen so that, in addition to their principal functions as light reflectors and the visible ceiling surface, they are reasonably effective panel absorbers of low- and mid-frequency sound.

A more careful use of descriptive terms is also imperative if we are not to delude ourselves when we specify materials. Many so-called "resilient" floor coverings, for example, are resilient only in comparison with concrete. Some of the plastic tiles are so hard and unyielding that they do little to cushion footfalls or impact sounds. Good carpeting, particularly over a good pad, provides excellent impact isolation as well as moderately high sound absorption; its use might obviate the need for acoustical tile or other absorbent material, and it might permit the use of lighter, more economical floor construction.

A reticulated urethane foam, glued to the inside of a cabinet panel of an accounting machine could provide high panel damping and high sound absorption within the cabinet. Thus, a little preventive effort at the source might make unnecessary an extensive and expensive treatment of the surfaces of the room in which the machine is operated.

Even the piercing scream of subway car wheels as the train rounds a sharp curve has been eliminated in one installation by the ingenious use of constrained damping on the wheels themselves.

As design and use of materials becomes more sophisticated, it is possible to provide very high sound and vibration attenuation with a minimum of added weight. The construction of passenger aircraft cabins is an excellent example of this. The cost per unit area for aircraft cabin fuselage construction is far greater than that for comparable attenuation in a skyscraper; but then, buildings need not fly. So weight reduction may actually be worth far more than the added cost of the materials often required for highly specialized sound barrier construction.

Sound control, then, is not simply a form of treatment, something added or attached to correct a problem. The existence of the problem may indicate a basic design deficiency. Even the words "design" and "control" suggest the possibility of preplanning and directing the evolution of the end product; in sound and vibration control this is certainly true. It is invariably far more simple, more economical, and more effective to avoid acoustical problems than to correct them after they appear.

ACKNOWLEDGMENTS

It is difficult to acknowledge properly my debt to the immediate contributors—intentional or unintentional—to this chapter. In more than a quarter of a century, one assembles, assimilates, and uses data from so many sources that it is impossible to be sure what came from where or whom. On the premise that most originality is undetected plagiarism, I wish to thank all of my colleagues and associates from whom I learned and accumulated the information on which this chapter is based.

I am particularly grateful to two of the great pioneers in the field—the late Dr. Paul Sabine and the late Paul Geiger. Working without the benefit of today's sophisticated equipment and advances in materials sciences, they took the first great steps that made possible much of what is done today.

To my former associates during my many years as an officer or committee member of the Acoustical Materials Association, I wish to extend sincere thanks for their generous contributions of information and advice. Manufacturers of acoustical materials have also been extremely cooperative and helpful.

REFERENCES

1. Lord Rayleigh, "The Theory of Sound." Dover, New York, 1945.
2. W. C. Sabine, "Collected Papers on Acoustics." Harvard Univ. Press, Cambridge, Massachusetts, 1922.
3. E. Buckingham, *Nat. Bur. Std. Sci. Papers* **20**, 193 (1925).
4. B. Kurtze and B. Watters, New wall design for high transmission loss or high damping. *J. Acoust. Soc. Am.,* **31** (1959).
5. C. Zwikker and C. Kosten, "Sound Absorbing Materials." Elsevier, Amsterdam, 1949.
6. S. Labate, Porous materials for noise control. *Noise Control* **2** (1956).
7. R. H. Bolt, On the design of perforated facings for acoustic materials. *J. Acoust. Soc. Am.* **19** (1947).
8. Acoustical Materials Association, "Performance Data on Architectural Acoustical Materials," Ann. Publ. Acoustical Materials Assoc., New York.
9. L. F. Yerges, Symposium on Acoustical Materials. Am. Soc. Testing and Materials *Spec. Tech. Publ.* **123** (1952).
10. L. Beranek and R. Sleeper, The design and construction of anechoic sound chambers. *J. Acoust. Soc. Am.* **18** (1946).
11. H. F. Olson, "Acoustical Engineering." Van Nostrand, Brooklyn, 1957.
12. L. L. Beranek, "Music, Acoustics, and Architecture." Wiley, New York, 1962.

13. V. Knudsen and C. Harris, "Acoustical Designing in Architecture." Wiley, New York, 1950.
14. T. J. Schultz, "Design Procedure for the Sound Absorption of Resonant Plywood Panels." Hardwood Plywood Inst., Arlington, Virginia, 1962.
15. L. L. Beranek (ed.), "Noise Reduction." McGraw-Hill, New York, 1960.
16. C. M. Harris (ed.), "Handbook of Noise Control." McGraw-Hill, New York, 1957.
17. D. Ross, E. Kerwin, and E. Ungar, Damping of plate flexural vibrations by means of viscoelastic laminae. In "Structural Damping" (J. E. Ruzicka, ed.), Am. Soc. Mech. Eng., New York, 1959.

GENERAL REFERENCES

1. T. Kinsler and A. Frey, "Fundamentals of Acoustics." Wiley, New York, 1962.
2. C. Harris and C. Crede (eds.), "Shock and Vibration Handbook." McGraw-Hill, New York, 1961.
3. J. Den Hartog, "Mechanical Vibrations." McGraw-Hill, New York, 1956.
4. S. Timoshenko, "Vibration Problems in Engineering." Van Nostrand, Princeton, New Jersey, 1955.
5. C. E. Crede, "Vibration and Shock Isolation." Wiley, New York, 1951.
6. W. P. Mason (ed.), "Physical Acoustics," Vols. 1–4. Academic Press, New York, 1964–1970.
7. "Sound: Its Uses and Control," Vols. 1 and 2. Acoustical Soc. Am., New York, New York, 1962–1963.
8. R. W. B. Stephens, "Acoustics and Vibrational Physics," 2nd ed. Edward Arnold, London, 1966.

MATERIALS PRODUCED BY ELECTRICAL DISCHARGES

Bell Telephone Laboratories, Murray Hill, New Jersey

I. Introduction

An electrical discharge is created in a gas when a sufficient number of electrons are removed from the atoms to make the gas an electrical conductor. A discharge can be generated by a variety of methods, but by each method enough energy must be provided to ionize the gas. All electrical discharges are characterized by the presence of both free charges and an electric field. The free charges are primarily electrons and positive ions, although negative ions often also exist. The region of the discharge where there is electrical neutrality among electrons and ions is called the plasma. Because of the high temperatures attainable and the chemical reactivity of species in the plasma, there are many applications of electrical discharges in the production of materials.

Two of the most obvious examples of electrical discharges are

[1] Formerly Faculty, Engineering Materials Laboratory, University of Maryland, College Park, Maryland.

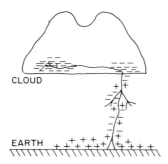

Fig. 1. Discharge of a lightning spark (1).

luminous neon signs and lightning. Lightning is a spark discharge. It is unstable and discontinuous, but it is still characterized by the flow of charge carriers in an electric field. If charged water molecules in a cloud make it a negative electrode relative to the earth, a discharge can occur if the potential gradient is sufficient for breakdown of the gas. As shown in Fig. 1, electrons are emitted from the cloud cathode and ionize the atmospheric gases by collisions as they move toward the earth anode. Photons produced in the initial ionization process can cause additional ionization. Therefore the degree of ionization varies along the path of the discharge, with the density of positive ions in the range 10^{11}–10^{19} per cm^3. The process of ion-electron recombination emits visible light, with intensity depending on the position in the discharge. For a cloud that is 500 meters above the earth's surface the potential difference can easily be 10^8 V. Although the time for the discharge of a single lightning spark is only several microseconds, the current can be over 10^5 A. Thus, the energy generated during a thunderstorm is often very destructive. Con-

TABLE I

Applications of Electrical Discharges in Producing Materials

Thermal discharges	Nonthermal discharges
Crystal growth	Chemical synthesis
Cutting	Metal compounds
Extractive metallurgy	Polymers
Machining	Coatings
Melting	Thin film microelectronics
Powder formation	Zone melting
Purification	Surface cleaning
Shaping	
Sintering	
Spray coating	
Spray forming	
Welding	

trol of this energy in artificially produced discharges makes possible myriad applications in science and technology. Many of the applications in the production of materials are listed in Table I.

A. HISTORICAL DEVELOPMENT

Although the electrical nature of the lightning discharge was noted by Benjamin Franklin in the middle of the 18th century, the energy available in an electrical discharge was not used to produce materials until the 20th century. In the 1920's, devices were developed in which an arc discharge could be created in a high-velocity stream of water vapor passing between carbon electrodes. The jet of ionized steam transferred thermal energy when it impinged on a surface placed in its path. These original devices suggested applications in metallurgical processes such as cutting and welding, but were not satisfactory because of contamination of the surface by vaporized electrode material and by the steam. In the following decades, further developments in laboratories of many large industrial organizations in the United States led to the use of gas-fed plasma torches and nonconsumable electrodes. Arc furnaces and arc welding equipment have been common for many years, but most of the other applications of the thermal energy of electrical discharges listed in Table I were not realized until recently.

The use of electrical discharges to produce materials by means other than their thermal energy also developed by evolution. Even before the 20th century it was known that bombardment of the discharge electrodes by heavy gas ions can eject atoms from the target. The electrode material can then be deposited on another surface placed in the discharge, or on the walls of the confining chamber. This process, called sputtering, occurs at relatively low gas pressures and in relatively diffuse discharges, as compared to the locally directed plasma jet. Contamination of the sputtered material by impurities present in the gas, and the low rate of deposition, prevented this method from competing effectively with the alternatives of thermal evaporation and electrodeposition. However, the development of high-vacuum equipment, and the demand for high-purity conductors and insulators by the solid-state electronics industry in the 1960's has made sputtering the most versatile method available to deposit coatings and thin films. Even more recently, new developments have taken advantage of the reactivity of species in electrical discharges to produce both metallic and nonmetallic materials with controlled properties and composition. This is now an active area of research.

Figures 2 and 3 show contemporary commercial processes for machining and metal shaping. In these processes the discharge of a series of capacitors is used to produce a low-voltage, high-current spark discharge.

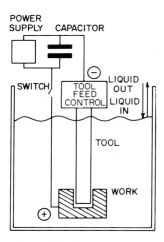

FIG. 2. Electrical discharge machining (2).

In electrical discharge machining (EDM) the workpiece is the anode and the tool is the cathode. By repetitive spark generation, with the discharge covering the entire face of the tool, the shape of the tool can be reproduced in the workpiece. It is a fact that the thermal energy produced in the spark machining process is sufficient to evaporate any material that is a good enough conductor to be made the anode. Although material can be machined at only about 10 cm³/min, the rate is independent of the hardness of the material. Therefore, EDM has an advantage for many aerospace metals, alloys, and even for cermets. Machining rates can be increased somewhat by increasing the power input, but this can

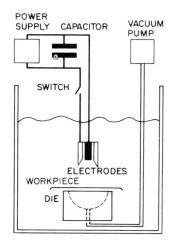

FIG. 3. Electrical discharge metal forming (3).

lead to undesirable changes in surface properties (4). The fluid covering the workpiece not only removes resolidified material, but also participates in the mechanism of the interaction between the plasma and the anode. In the electrical discharge forming process, the liquid transmits the energy from the discharge and forces the material to take the shape of the female die, as shown in Fig. 3.

Although the range of applications of EDM are now realized, a long period of time elapsed before EDM and other more versatile applications of electrical discharges were used in producing materials. There are good reasons for this. In addition to the practical difficulties discussed above as encountered with plasma jets and sputtering, there was a lack of understanding of the complex phenomena that occur in electrical discharges, and a lack of equipment to generate high-temperature discharges. These prevented the use of such discharges in materials technology. It is fortunate that within the last 20 years the physicist's interest in plasmas, the demand by aerospace industries for high-temperature materials, and the need for high purity in electronic components have found a common ground. Today the equipment for generating various types of discharges for use in producing materials is commercially available. Plasma research extends into areas far beyond the applications in materials, however.

B. PROPERTIES OF IONIZED GASES AND GENERAL APPLICATIONS

The positively charged ions and negatively charged electrons in a gas discharge are accelerated in opposite directions by the electric field. The electrons, due to their smaller mass, acquire a considerably higher velocity than the ions. The temperature equivalent of the kinetic energy acquired by passage through a potential drop of 1 eV is in the order of magnitude of $10^{4}°$K. This energy is transferred to the other particles in the plasma by collisions, and therefore the entire gas can be heated. Although the amount of energy transferred during each collision is small, at relatively high pressures the frequency of collisions may be sufficient to distribute the thermal energy uniformly among the electrons, ions, and un-ionized species. High-pressure plasmas, such as the one shown in Fig. 4, operating at approximately atmospheric pressure, possess the thermal properties required for applications in cutting, spray coating, and ultrafine particle formation.

On the other hand, for discharges in a pressure range below several Torr, commonly used for low-temperature applications such as sputtering and chemical synthesis, the electron and ion temperatures can differ by several orders of magnitude. Figure 5 illustrates a low-pressure discharge, referred to subsequently as a nonthermal plasma. The gas is usually less than 1% ionized.

Although the frequency of collisions increases as the gas density is

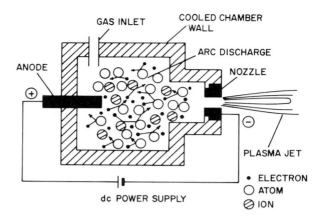

FIG. 4. Schematic drawing of an arc plasma torch. Usual operating pressure is approximately 1 atm.

raised, in general, thermodynamic equilibrium among electrons, ions, and neutral particles is never completely reached in the plasma. It is reasonable to expect that as the energy input is increased, the degree of ionization of the gas and its temperature will also increase, and conditions of local equilibrium may be approached. If the particles in a relatively high density plasma are assumed to have a Maxwellian velocity distribution, then the Saha equation can be used to calculate the temperature of the plasma as a function of the degree of ionization (5). In argon, for example, the ionization equilibrium reaction is

$$A + E_I \leftrightarrows A^+ + e \tag{1}$$

and the Saha equation is

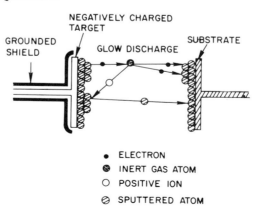

FIG. 5. Glow discharge sputtering in a low-pressure discharge.

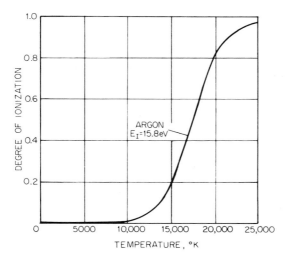

FIG. 6. Temperature plotted against degree of ionization calculated from the Saha equation (5).

$$\frac{N_e N_{A^+}}{N_A} = \frac{N_{A^{2+}}}{N_A} = \text{(constant)} \; T^{5/2} \exp(-E_I/kT) \qquad (2)$$

where E_I is the ionization energy of the argon atom (15.7 eV = 361 kcal/mole), and N_e, N_A, and N_{A^+} are the densities of electrons, atoms, and ions, respectively. Figure 6 is a plot of temperature versus degree of ionization for an argon plasma at 1 atm pressure. The temperature depends on the particular gas, since the ionization energy is different for each, as shown in Table II. The choice of the proper gas is dictated by the particular application and by economics. Reactive gases such as

TABLE II
IONIZATION POTENTIAL OF GASES

Gas	E_I(eV)
Argon	15.7
Helium	24.6
Hydrogen atom	13.6
Hydrogen molecule	15.4
Krypton	13.9
Methane	14.5
Neon	21.6
Nitrogen molecule	15.8
Oxygen atom	13.6
Oxygen molecule	13.0
Xenon	12.1

hydrogen, nitrogen, and oxygen may be chosen intentionally for some applications, whereas great pains may be taken to completely eliminate them for other applications.

1. Thermal Plasmas

The thermal plasma is the only practical means of maintaining temperatures from 5000°K to 40,000°K for extended periods of time. Applications in the production of materials usually utilize the temperature capability in the 10^3°K range, since this is sufficient to melt or evaporate any solid that does not decompose. In comparison with other sources of heat, the gas plasma gives the highest temperature and the highest heat transfer rate and is often the most economical. In devices such as the one shown in Fig. 4, the temperature and heat transfer rate depend on the velocity of the gas. Even though heat losses may be 50% or more, the operating cost is comparable to oxygen-hydrogen flames. However, the initial equipment costs are higher. Various heat sources are compared in Table III. In the plasma, as opposed to chemical flames, there is no limitation due to theoretical flame temperatures that depend on the heat of reaction of a fuel. In principle any gas can be used to generate the plasma.

TABLE III
COMPARISON OF HEAT SOURCES (6,7)

Heat source	Maximum temperature, °C	Gas velocity, ft/sec	Heat transfer rate, Btu/in² sec
Induction furnace	4,000		
Resistance heating	3,000		
Solar furnace	3,000		
Flames			
Methane ($CH_4 + O_2$)	1,650	20	0.5
Acetylene ($C_2H_2 + O_2$)	3,100	400	5
Hydrogen ($H_2 + O_2$)	3,500		
Cyanogen ($C_2N + O_2$)	4,850		
Carbon subnitride ($C_4N_2 + O_2$)	6,000		
Welding arc	11,000	20	70
Plasma arc torch	22,000	15,000	250

2. Nonthermal Plasmas

In low-pressure discharges used to deposit materials by sputtering, the impact of ions accelerated toward the negatively charged target causes atoms of the target to be removed. This is illustrated in Fig. 5. Material is ejected from the surface as single atoms or small groups of atoms, as opposed to the bulk melting or evaporation that is produced by thermal plasmas. In fact, the target and substrate are often artificially cooled, since the ion energy that is dissipated as heat can lead to undesirable effects on the structure of the deposited material.

Low pressures are required in sputtering applications to minimize collisions between sputtered atoms and gas-phase impurities during the transport of material from the high-purity target to the substrate. If the distance between the target and substrate is kept as small as possible, these undesirable reactions and the return of sputtered atoms to the target can be reduced. Thus, the lowest pressure at which the discharge can be sustained, commensurate with reasonable sputtering rates, is most desirable. Since the sputtering phenomena take place at the electrodes, the plasma region of the discharge functions only as a conduction path for the ions.

A minimum ion-energy characteristic of the target material is required to sputter an atom. This threshold energy is related to the sublimation energy of the material, and is therefore in the range 5–25 eV (115–575 kcal). Above this threshold energy the yield of atoms sputtered per incident ion is usually less than one, and varies with the ion energy, angle of incidence, and crystallographic orientation of the target (8). In the process of physical sputtering that is shown in Fig. 5, heavy inert-gas ions increase the sputtering rate as compared to lighter ions. Although a complete understanding of the mechanism of sputtering has not yet been obtained, it is known that sputtered atoms are ejected preferentially in certain crystallographic directions (9). Several atomic layers in the target material may therefore be involved.

Since sputtered atoms are ejected with an appreciable energy compared to thermally evaporated material, films deposited by sputtering have superior adhesion to the substrate. In addition, in sputtering there is no crucible contamination, so that refractory materials can be handled without difficulty. The stoichiometry of alloys can be preserved by sputtering. Generally this is very difficult to do by evaporation or electrochemical deposition. Sputtered films usually possess good thermal stability, uniformity in thickness, and reproducible electrical properties. There is greater flexibility in deposition of metals, alloys, semiconductors, and dielectrics, irrespective of their melting temperatures.

In some applications, such as the sputtering of bismuth in hydrogen discharges, direct removal of the metal does not occur as described above (10). In this chemical sputtering a volatile compound is formed on the target. This compound vaporizes and subsequently decomposes in the plasma. The pure metal is then deposited on the substrate. Alternatively, the reactivity of species in the plasma can be used to advantage in a process called reactive sputtering (11). A metal target is sputtered into an electrical discharge of a gas such as nitrogen or oxygen, forming oxides or nitrides of the metal that are deposited on the substrate. By varying the composition of the gas during deposition it is possible to alter the stoichiometry of the film. This technique has found recent use in the production of graded polymer coatings (12) and barium titanate ferroelectrics (11).

The reactivity of the particles in the plasma state can also be used in other types of chemical syntheses. Because activated states are produced in the gas phase, some processes are competitive with alternative catalytic synthesis. Most of the recent interest and reporting (13,14) has been in the formation of hydrocarbons and other compounds that either are vapors under normal conditions or have high vapor pressures compared to metals. However, only solid products are considered in this chapter. Applications in which metals (15), polymers (16), and even refractory compounds (17) are produced by gas-phase reactions in electrical discharges are described in Section III.

Although it is clear from the above discussion that the high temperature and the reactivity of ionized gases have many practical applications in the production of materials, the major difficulty is to generate, maintain, and control the conditions in the electrical discharge. Since the materials used may be in bulk forms, powders, coatings, or thin films, the techniques employed vary.

II. Methods of Generating Discharges

Energy must be added to a gas to create the electric field and the charge carriers essential to any electrical discharge. Several methods are available to supply this energy. A very direct, nonelectrical method of ionizing a gas such as argon utilizes radiation of wavelength corresponding to the ionization energy. The appropriate wavelength is calculated from Eq. (3) with the values of Planck's constant, the velocity of light, and the ionization energy:

$$\lambda = hc/E_I = \frac{(6.62 \times 10^{-27}\ \mathrm{erg/sec})(3 \times 10^{10}\ \mathrm{cm/sec})}{(15.7\ \mathrm{eV})(1.6 \times 10^{-12}\ \mathrm{erg/eV})} = 7.91 \times 10^{-4}\ \mathrm{cm}$$

$$(3)$$

Radiant energy sources are impractical for sustained discharges, and only electrical sources are useful to produce discharges for materials applications. However, these methods also cause changes in electron energy levels during excitation, ionization, or recombination in the gas. This gives rise to photon emission, which causes the luminous zones characteristic of most electrical discharges. This is always observed in the arcs and glow discharges used to produce materials; the color of the discharge naturally depends on the particular gas involved.

The electric field can be applied with either a dc or an ac source. The latter may or may not require electrodes. The mechanism of generation and the properties of the discharge are determined by the method used to supply the electrical energy. For this reason equipment is designed to take advantage of the properties of each type of discharge.

A. Types of Discharges

1. dc

If the electric field is generated between electrodes, with no external electron source, such as a thermionic emitter, then the current follows the curve shown schematically in Fig. 7, as the voltage is increased. This curve defines several types of electrical discharges. In general the electric field strength and the pressure of the gas determine the type of discharge generated.

At very low voltages the few free electrons and ions that exist in an ordinary gas under ambient conditions are accelerated to the electrodes. These charged species are neutralized at the electrode surfaces to produce the small currents observed. Eventually the current levels off, and

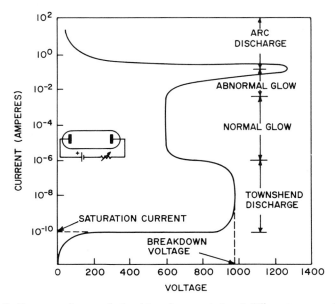

Fig. 7. Current-voltage relationship, characteristic of different types of electrical discharges.

the saturation value is limited by the number of charge carriers that can reach the electrodes. A further increase in voltage gives some electrons sufficient energy to ionize gas atoms, thereby generating additional charged particles. In addition to ionization, there is a current increase due to secondary electron emission from the cathode as it is bombarded with positive ions. The current increase leads to breakdown in the gas, as indicated in Fig. 7.

The Townshend discharge that occurs before breakdown operates with a current of approximately one microampere, and is of no practical value in materials applications. The ion energy is too low for sputtering, and the temperature of the plasma region is not high enough to be used effectively for melting or evaporation.

Following breakdown in the gas an increase in the applied voltage generates at least as many electrons by secondary emission and ionization as are lost at the anode or by recombination in the gas. In this condition the discharge is referred to as self-sustaining, and the sharp rise in current causes a transition to either an arc or a glow discharge. The arc discharge usually occurs when the pressure is below several Torr. It can be stabilized by controlling the current with a resistance placed in series with the dc source. Since the pressure is low, and processes that take place at the electrodes sustain the glow discharge, this type of discharge is obviously important in sputtering applications. The presence of the secondary electron emission mechanism that maintains a stable glow discharge is the reason why the plasma region is not essential, as mentioned previously in Section I,B.

A glow discharge is characterized by currents in the milliampere range, and by distinct luminous regions caused by space charges near the electrodes. There is a negative electron space charge near the anode, and a positive ion space charge near the cathode. Most of the potential drop in the glow discharge is at the cathode, since the electron mobility is much higher than the mobility of positive ions. In the stable glow discharge region of Fig. 7, the discharge covers only part of the cathode. Therefore, as the voltage is increased the current density is permitted to remain constant by an increase in the area of the discharge. This continues until the entire electrode surface is covered. Then the current density again begins to increase, and the discharge region is referred to as the abnormal glow. A large area of coverage is desirable in sputtering applications, but care must be taken to isolate the back of the electrode so that the discharge occurs only on the front side. This is usually accomplished by placing a grounded shield within the cathode space charge region, as illustrated in Fig. 5.

A glow discharge can be operated at a pressure as high as 1 atm, but ordinarily above a few centimeters of mercury the current increases rapidly owing to heating of the cathode, and the glow discharge is transformed into an arc. The arc discharge is also self-sustaining, but in contrast to the glow discharge mechanism, electrons are emitted from the cathode by thermionic emission or field emission rather than by ion bombardment. In the arc discharge the current is very high, and the voltage is low, as shown in Fig. 7. Arcs operating at atmospheric pressure have the high temperatures needed in many applications. In a flowing gas, or arc plasma torch, convection currents distribute the heat content and intense luminosity throughout the gas.

Thermionic arcs are most common in materials applications. In practice, the arc discharge is generated without passing through the glow discharge region. If breakdown is made to occur locally between the electrodes, a transient spark can give rise directly to a stable arc, or, if the electrodes are put in contact, the high current flow can be maintained in the form of an arc when they are separated.

Another type of electrical discharge, one that has been used in chemical synthesis, is the corona. This discharge is usually generated between cylindrical electrodes, and differs from an ordinary glow discharge in that breakdown occurs in a nonuniform field. Current in the milliampere range can be sustained in a corona discharge. Although the mechanism of generation of the corona is similar to that of the glow discharge, it is not nearly as important in the applications discussed in Section III.

2. ac

Arcs and glow discharges can be generated with an alternating field applied between the electrodes shown in Fig. 5. Alternatively, the ac discharge can be generated without electrodes. For thermal and nonthermal applications of ac discharges in producing materials, the electrodeless discharge operating with a source of electromagnetic radiation at frequencies above 1 Mc/sec is most important. At frequencies in the kc/sec range, ac and dc discharges are both established by essentially the same mechanism. However, in the high-frequency electrodeless discharge there obviously can be no space charge phenomena at an electrode. The entire discharge is then similar to the plasma region of a dc discharge.

In the radio-frequency and microwave frequency ranges (10^6–10^{11} cps) the mechanism of generating a self-sustained discharge differs considerably from the mechanism discussed in the previous section. The processes that initiate the discharge with an ac source depend on the frequency, the pressure of the gas, and the dimensions of the enclosure. At radio frequencies the energy from the ac field can be coupled directly to the electrons in the gas with an rf induction coil wrapped around the discharge container. In the microwave region, above approximately 10^9 cps, coupling is obtained by placing the container in a waveguide. The processes that occur as the frequency is increased are described briefly below to illustrate the mechanism operating in high-frequency discharges and their advantages in certain applications. The detailed theory of microwave breakdown in gases and the general applications of microwaves in heating have been described in recent books by MacDonald (18) and by Puschner (19).

At frequencies below 1 kc/sec, charged particles can traverse the discharge chamber before the direction of the field reverses. If there is no external electron source such as a heated filament to initiate the discharge, breakdown and current multiplication occur by ion collisions with the electrodes or with the chamber walls. The current-voltage characteristic is the same as in Fig. 7. As the frequency is increased electrons begin to oscillate in the alternating field and do not reach the electrodes or the boundaries of the chamber. Compared to the dc discharge, at megacycle frequencies the electron loss at the electrodes is low. However, the breakdown voltage does not necessarily decrease since there is no secondary electron emission to supplement the current. With further increase in frequency a point is reached at which no loss occurs at the electrodes or walls, and the large current multiplication by collisions in the gas does lead to a decrease in the breakdown voltage. Therefore, in this frequency range the sustained current is due only to excitation and ionization from interactions among atoms, electrons, and ions. High-frequency discharges operate at lower total voltages than dc discharges generated in the same vessel.

High-frequency discharges are used in nonthermal applications to deposit materials by sputtering or by reactions of gas-phase compounds in the plasma. Discharges can be sustained in the pressure range 10^{-3}–10^{-2} Torr, although at the lower pressures charged particles can reach the chamber walls. A magnetic field superimposed on the ac field changes the electron motion, causing electrons to be accelerated in a spiral path. This enables even lower pressures to be used without wall effects.

One important reason for the use of rf discharges in nonthermal applications is that insulators cannot be sputtered with a conventional dc potential applied to the electrodes. With insulators the dc source cannot continuously accelerate positive ions to the surface. The positive charge builds up and prevents sputtering. However, if a dielectric target is placed in front of an rf electrode, the positive charge buildup can be neutralized on each negative half-cycle of the alternating source. At frequencies above 10 kc/sec a net negative bias occurs on the dielectric because of the higher mobility of electrons compared to ions, and continuous sputtering can be performed.

In the synthesis of materials in electrical discharges the electrodeless discharge has several advantages. Electrodes can be contaminated by the deposition of the product, and the field gradients that occur near electrodes can lead to nonuniform products. These factors are especially important in polymer deposition. In high-pressure as well as low-pressure applications, electrodeless discharges are useful where reactive gases such as oxygen and nitrogen are present. These gases tend to corrode electrode surfaces. This is the major problem that has led to the development of the rf plasma arc torch used for spray coating and crystal growth.

B. EQUIPMENT AND SYSTEMS FOR PRODUCING MATERIALS

Whether the electrical discharge is used to melt refractory materials in a high-temperature arc, or to deposit thin polymer films in a glow discharge, certain features are basic to all systems. A power source to generate and maintain the discharge, a supply of gas and reactants in controlled amounts, and a method of localizing the discharge to the area where the material is to be produced are a few of these features. The equipment required and the operating conditions depend on the particular material, but a few of the most common systems are described below.

1. Arc Plasma Torches

When gases are passed through an electric arc they absorb energy and can therefore be heated to high temperatures without the combustion reactions that occur in flame torches. The latter have been described in detail by Ault and Wheildon (20). Arc plasma torches can be constructed to operate with either dc or ac power sources. In the simplest type of dc plasma torch, an arc discharge is generated between a solid metal electrode fixed at one end of a chamber, and another electrode with an opening, or nozzle, at the other end. This is shown schematically in Fig. 4. A plasma jet is formed by the gases that are ejected from the nozzle. This flowing gas can be used to melt or to vaporize material that it impinges upon, or powder that is injected into the gas stream can be melted and redeposited. Therefore, the plasma torch can be used in chemical synthesis, crystal growth, extractive metallurgy, and spray coating.

The nozzle is usually cooled to prevent rapid destruction. However, cooling of the outer region of the discharge, as well as the electrodes, has a more important effect that is utilized in modern plasma torches, such as the plasma spraying unit in Fig. 8. In devices of this kind intentional cooling of the periphery of the arc leads to a decrease in its diameter and to a subsequent increase in its axial temperature. This constriction effect is referred to as the "thermal pinch." Cooling of the

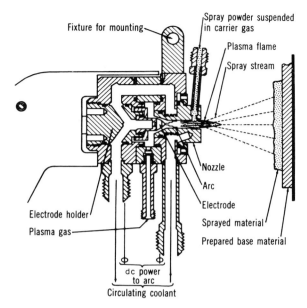

FIG. 8. A modern arc plasma torch for powder spraying (22).

outer region causes the electrons to be attracted to the axis of the chamber. Constricted arcs, based on external cooling, provide increased stability, higher axial temperature, and higher gas velocity through the nozzle. An external magnetic field can be used to further increase the amount of constriction in arc plasma devices. However, in most modern plasma jet devices sold commercially this "magnetic pinch" effect is not used.

In the plasma spraying torch in Fig. 8, the arc is confined between the two electrodes. In some applications, however, it is desirable to extend the arc beyond the nozzle. Devices of this type are used in plasma cutting and in the vaporization of powdered refractory ores (21). In either type of torch the gas flow pattern has a large effect on the arc behavior, and cooling the periphery is desirable. Cool gas may be added to the torch tangentially to create a vortex within the discharge, or it may be added axially along the walls of the chamber. Both methods generate the thermal pinch effect. The feed lines for cooling the anode nozzle, for adding the working gas, and for adding power are made small enough so that the torch can be held in the operator's hand. A tungsten-2% thoria cathode at the rear of the chamber and a copper anode at the front provide stable operation. With an inert gas the axial temperature can be continually maintained at about 14,000°K under a wide range of operating conditions without electrode replacement for as many as 100 hours. With only a change in the outlet nozzle, the plasma torch can be used with many gases, including hydrocarbons, although electrode erosion increases with the reactivity of the gas.

A complete system designed for the specific purpose of plasma spraying is shown in Fig. 9. Most systems of this type operate with a maximum power input of 20–40 kW, and produce arc plasma currents up to several hundred amperes. The power source should have a falling voltage-ampere characteristic. The gas flow rate varies in particular applications, and depends on the nozzle design, but typical values are between 25 and 200 cu ft/hr. In general, for maximum heat transfer to the gas, low flow rates are preferred. However, it is the high flow rate that gives the material injected into the plasma jet the momentum required to produce adherent deposits.

Plasma torches using electrodeless discharges are operated with rf power sources, and have several advantages and disadvantages compared to devices that use dc sources. There are no limitations on the temperature or the reactivity of the gas with an rf torch, since there are no electrodes that can be eroded. The arc discharge is usually generated in a quartz tube placed within the rf coil, as in Fig. 10. One end of the tube is open, and the other end usually contains a removable thermionic electrode that is heated inductively to a sufficiently high temperature to start the discharge. The radius of the discharge chamber is limited, since most of the energy from the rf source is dissipated near the walls. Even with this skin-effect limitation, the highest temperature is obtained along the tube axis if the walls are cooled, as in the constricted dc plasma torch. Tangential flow of the cool incoming gas also increases

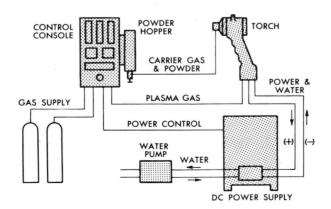

Fig. 9. A complete system for plasma spraying (23).

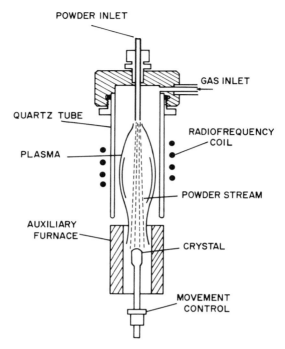

FIG. 10. An electrodeless plasma torch used for crystal growing (24,25).

the stability of the arc. With a typical rf power supply operated from 1 to 25 Mc/sec, with a peak power of 10 kW, the gas flow rate for most applications is regulated between 15 and 60 cu ft/hr. The longer residence time of the gas in the arc chamber makes the rf torch more suitable for chemical synthesis than for depositing adherent plasma-sprayed coatings. The absence of electrodes in the rf torch makes it easy to feed the starting materials directly along the axis of the chamber.

The rf plasma torch in Fig. 10 is used to advantage in the growth of single crystals of high-melting-point materials by the Verneuil method. Polycrystalline material is injected into the arc and is melted and resolidified on the growing crystal. A molten droplet is usually maintained on the top of the crystal, and this continuously solidifies as the seed is lowered and rotated. The auxiliary furnace is used to reduce the temperature gradient in the growing crystal. Many of the other applications of the rf plasma torch are discussed in greater detail in Section III.

2. Glow Discharge Deposition

The problem of producing uniform, high-purity materials with controlled properties has been approached from several directions. General applications and procedures in vapor deposition have been described in a recent book by Powell, Oxley, and Blocher (26). In sputtering, it has been possible through high-vacuum equipment and the proper orientation of the substrate with respect to the glow discharge electrodes to control the deposition rate, adhesion, purity, and thickness satisfactorily in the formation of both conducting and insulating layers.

a. Diode Sputtering

Vacuum systems using modular components have been developed to provide maximum flexibility. The system shown in Fig. 11a for sputtering between two parallel electrodes uses a feed-through collar at the base of a bell jar to introduce the required leads. High-voltage leads are connected between the cathode target and the anode substrate holder.

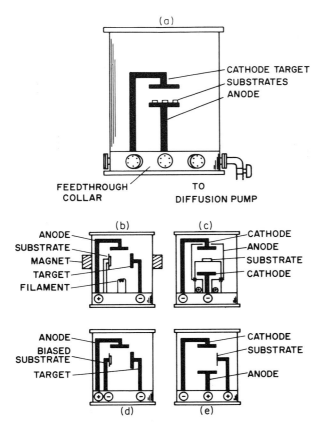

FIG. 11. Variations in glow discharge sputtering techniques: (a), diode sputtering, (b), triode sputtering, (c), getter sputtering, (d), bias sputtering, (e) plasma anodization.

Power sources that operate at 1 to 5 kV and 300 mA are suitable for the dc sputtering of most materials.

Although insulators can be sputtered only with an alternating field, for conductors rf sputtering can also be used. However, it is then necessary to block the flow of dc by placing a capacitor between the high-voltage rf power source and the electrode. Power requirements for rf sputtering are similar to those for dc, but a careful choice of the frequency and of the technique for coupling the source to the sputtering chamber is needed to avoid undesirable dissipation of the power as heat. The Federal Communications Commission does not fix the frequency that must be used, but it does enforce regulations to prevent industrial rf power equipment from interfering with radio communications. To use the rf power efficiently for sputtering, the impedance of the glow discharge and the electrodes must be matched to the output of the rf generator.

Even though diode sputtering is done in the pressure range 10^{-3} to 10^{-1} Torr, a diffusion pump with an ultimate pressure of 10^{-6} Torr is used to clean adsorbed gases off the chamber walls, and to minimize pumpdown time. A micrometer valve in the feed-through collar is desirable so that sputtering can be carried out using a dynamic argon flow, balanced by continuous pumping of the system. This is important, since the chamber walls can outgas when the discharge is initiated. The residual gases in the chamber are primarily oxygen, nitrogen, and water vapor. These gases can contaminate sputtered metal films if their partial pressures are not kept low. On the other hand, reactive sputtering can be carried out intentionally, to deposit oxides, for example, by adding a premixed gas containing oxygen, or by placing a separate leak valve on the feed-through collar. Other feed-through openings are used for a pressure gauge, masks, a substrate heater, and a water source for cooling the cathode.

Although all metals can be deposited by cathode sputtering, the rates vary between 100 and 1000 Å/min. This depends on the target material, the applied voltage, the gas pressure, the distance between the target and substrate, and the shielding of the target perimeter.

b. Triode Sputtering

Several disadvantages of diode sputtering have led not only to the use of rf rather than dc power sources to lower the operating pressure, but also to changes in the electrode configuration. The alternative electrode configurations shown in Fig. 11 yield higher purity at increased deposition rates. In triode sputtering, Fig. 11b, a thermionic emitter is used as an external electron source. Since the probability of ionization when these electrons are attracted to the anode is increased, the discharge can now operate at a pressure of 1×10^{-3} Torr. If, in addition, a magnetic field is used to confine the discharge, pressure as low as 5×10^{-4} Torr can be used with success. The third electrode is the sputtering target. A negative potential of about 2 kV is applied to it to attract positive ions from the discharge. In triode sputtering, the substrate, placed opposite the target, is not at all involved in the process of generating the plasma. Although additional power supplies are needed, triode sputtering clearly has advantages compared to the conventional dc sputtering with only a cathode target and an anode substrate.

c. Getter Sputtering

Another method of keeping the pressure of residual gases low, called getter sputtering, is shown in Fig. 11c. The anode is enclosed in a container with only a small opening for pumping and backfilling with argon. Since the walls of the container are large in comparison to the area of the substrate, metal sputtered onto the walls reacts with most of the residual gases. By this technique the vessel can be operated at 10^{-2} Torr, but the partial pressure of oxygen, nitrogen, and water vapor may be as low as 10^{-10} Torr.

d. Bias Sputtering

To prevent adsorbed impurities from contaminating sputtered superconductor films during deposition, Maissel and Schaibel (27) applied a small negative bias to the substrate, as illustrated in Fig. 11d. In bias sputtering, positive ion bombardment continually removes material from the substrate, but at a lower rate than it is deposited. Adsorbed impurities are removed preferentially.

A process analogous to bias sputtering is assymetric ac sputtering (28), in which the ac potential between the target and the substrate causes sputtering from both surfaces alternately. However, if the current density is limited during periods when the substrate is negative, a net deposition of material with increased purity occurs on the substrate.

Although materials deposited by these two techniques have a high purity, their adherence to the substrate is inferior to that of films deposited without bias. However, for superconducting thin films the lack of contamination is most important.

e. Plasma Anodization

In addition to the ac methods for sputtering dielectrics, high-quality oxide films for integrated circuits can be produced by the technique of plasma anodization. As shown in Fig. 11e, a dc bias is applied to a conducting substrate, such as silicon or aluminum, placed in an oxygen glow discharge. Reactive species are extracted from the discharge. If the oxide film forms at the plasma-oxide interface, then transport of metal cations is required for continued growth. This field-induced transport is analogous to anodization in liquid electrolytes. Oxide films up to several thousand Angstroms have been grown on heated substrates (29). A dc, rf, or microwave discharge can be used.

f. Ion Plating

The adhesion of evaporated deposits can often be improved by placing the substrate in an inert gas discharge and evaporating the material into the discharge. The increased energy of the condensing atoms, combined with the cleaning of the substrate by ion bombardment, lead to the improved adhesion. Naturally, ion plating is limited to materials that can be evaporated by thermal or electron-beam techniques.

g. Chemical Synthesis

Solid products can be produced from vapor-phase reactants in an electrical discharge. High-melting-point materials, such as carbides, oxides, and nitrides, can be deposited on unheated substrates. This is especially important in the deposition of thin films for electronic applications. As an example, Fig. 12 shows a method to synthesize boron nitride by reacting boron trichloride with ammonia in an electrodeless discharge.

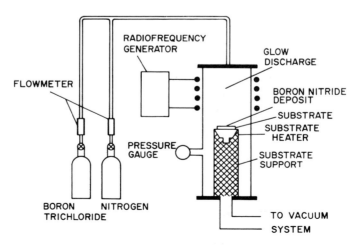

Fig. 12. A flow system for the synthesis of boron nitride in a radio-frequency discharge.

Thermal decomposition of the reactants would require a much higher temperature.

The species that exist in the discharge region, and the reactions that occur, may be very complex, but stoichiometric compounds can be deposited. Adsorption of reactants on the substrate, followed by nucleation and growth of the solid compound, are the atomic processes operating in the film formation. The heterogeneous surface reaction leads to more adherent deposits than if the products were formed in the gas phase. Since many gases are available in purer form than that of sintered sputtering targets, materials of higher purity can be produced by chemical synthesis.

h. Hollow Cathode Heating

If the cathode of a dc glow discharge is spherical, as shown in Fig. 13, then the electrons emitted from the cathode are accelerated by the potential field near the surface, and can be focused onto a specimen at the center of the sphere. In this new development in electron beam heating (30), no space charge is built up on insulators since the specimen is not part of the discharge circuit. A molten zone can therefore be formed in a variety of refractory materials, and purification and crystal growth by zone melting can be carried out. As long as a stable discharge can be maintained, atmospheres of inert, oxidizing, or reducing gases can be used.

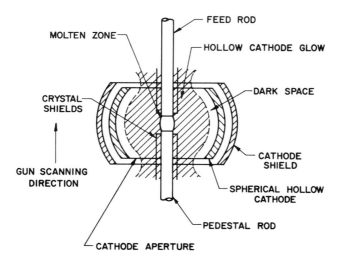

FIG. 13. Hollow-cathode heating used for zone refining and crystal growth (30).

III. Applications

A. ARC PLASMA SPRAYING

The general properties and applications of flame-sprayed coatings have been the subject of previous chapters in *Modern Materials* (20,31). Since the temperature attainable in the plasma arc torch is higher than in fuel-burning torches, coatings of even the most refractory materials can be formed. Tungsten [melting point (m.p.) 3410°C], with up to 98% of the theoretical density, and tantalum and hafnium carbides with melting points above 3800°C, can therefore be used as coatings in a variety of applications. For lower-melting-point materials, it is sometimes still advantageous to use plasma spraying even though the rate and efficiency is higher with more conventional techniques. Better mechanical bond strength and higher densities can be obtained because of the high velocity of the molten coating material in the plasma jet. For example, Fig. 14 compares deposition of beryllium oxide (m.p. 2515°C) with plasma and oxyacetylene torches. Similarly, deposits of zirconium oxide (m.p. 2600°C), which is difficult to heat because of its low thermal conductivity (32), are harder and have a higher density when plasma sprayed. The products of combustion in fuel-burning torches can contaminate coatings, but this is eliminated in plasma spraying if an inert gas is used. The lamellar structure of the deposited layer is characteristic of both types of coatings.

The applications in rocket engines and reentry vehicles are now well

FIG. 14. Beryllium oxide deposited on beryllium metal by spraying with an oxyacetylene torch (left) and by plasma spraying (right). The density of the plasma sprayed oxide is greater (22).

established, but less glamorous uses also exist, as in the formation of catalysts, radiation shields, heat sinks, and self-supporting parts. Also, inexpensive base materials can be coated to produce wear-resistant tools. Through these and other applications, arc plasma spraying is now established as a versatile technique for use when coatings resistant to heat, friction, and corrosion are required.

1. Sprayed Coatings

The proper operating conditions are important in plasma spraying. If powders are used as the source of material, their size distribution must be chosen so as to avoid vaporization of the smaller particles. From the data in Table IV it is evident that the high enthalpy of hydrogen is desirable for operating the plasma torch. However, hydrogen alone erodes the electrodes, and mixtures of 90% N_2 with 10% H_2 are usually used. The cost is certainly less than for an inert gas. Even with this mixture, undesirable effects are still possible, as in the examples of plasma-sprayed coatings discussed below.

TABLE IV

OPERATING CHARACTERISTICS OF PLASMA TORCH FOR SPRAYING
POWDERS (33)

Gas	Power, kW	Plasma temperature, °C	Enthalpy, Btu/lb	Arc voltage	Efficiency
N_2	60	7,300	17,900	65	60
H_2	62	5,100	138,000	120	80
He	50	20,000	92,000	47	48
A	48	14,500	8,400	40	40

In plasma spraying the crystal structure and chemical composition of the coating are often not the same as those of the powder fed to the torch. This is due to the atmosphere and temperature that the material is exposed to. Levinstein (33) found that hafnium and tantalum carbides sprayed as powders give more than one phase in the deposit. The X-ray diffraction patterns of the carbide coatings are poor because of strains in the crystal lattice, but after vacuum heat treatment, compounds of hafnium containing both nitrogen and carbon could be identified. In addition to TaC, Ta and Ta_2C appeared in the tantalum carbide coatings. Therefore, the high temperature of the plasma partially dissociates the starting materials, with a resultant loss of carbon. Table V shows the chemical composition before and after deposition. Unwanted TaN, HfO_2, and TaO_2 are also formed. Substitution of argon for nitrogen in the gas mixture eliminates the nitrides, but HfO_2 and Ta_2C are still found. Some uncombined Ta is still deposited also.

TABLE V

CHEMICAL ANALYSIS OF PLASMA SPRAYED CARBIDES (33)

Material	Chemical analysis, %			
	Metal	Carbon	Nitrogen	Oxygen
Hafnium carbide				
Before spraying	93.6	6.31	0	0
After spraying	94.1	2.92	0.058	1.88
Tantalum carbide				
Before spraying	93.8	6.23	0	0
After spraying	93.0	1.12	0.326	1.96

Tungsten and molybdenum coatings are very brittle. In general, heat treating of the deposits alters their properties. Table VI shows the changes in density, hardness, and tensile strength of tungsten and molybdenum powders deposited on aluminum by plasma spraying. Similar results for tungsten on copper have been obtained by Spitzig and Form (34). Even though recrystallization and grain growth occur during the heat treatment, the final properties of the coatings may still depend on the form in which the original material is fed to the torch.

Graded coatings can be built up to provide better thermal properties. For example, in the system used for the coatings shown in Fig. 14 it is possible to deposit beryllium at the base and gradually increase the proportion of BeO as the deposition progresses. This improves the bonding at the metal-coating interface, and minimizes the effects of thermal expansion during high temperature service. Coatings with variable porosity can be formed by changing the distance between the source and the substrate during deposition.

2. Spray Forming

By depositing material on a mandrel that can subsequently be dissolved, or otherwise removed, objects can be fabricated from refractory materials that are extremely difficult to machine by conventional methods. Moss and Young (32) describe the spray forming of tungsten parts with both internal and external threads. For items produced on a larger scale, such as crucibles, it is possible to spray materials on reusable mandrels. Hollow shapes can be made without difficulty, and masking of the mandrels allows the formation of objects with openings in their walls. Thus, plasma spray forming offers an alternative to powder sinter-

TABLE VI

PROPERTIES OF PLASMA-SPRAYED TUNGSTEN AND MOLYBDENUM (33)

Material	Density (% theoretical)		Vicker's hardness		Tensile strength, psi				
	As sprayed	Heat treated[a]	As sprayed	Heat treated[a]	75°F	2,000°F	3,000°F	3,500°F	4,500°F
Tungsten	83.0-84.0	84.5-85.5	330-390	162-200	6,700	15,300	7,100	3,800	1,600
Molybdenum	86.2-87.1	89.3-90.2	321-368	125-175	30,500	10,500	—	—	—

[a] Tungsten heated at 4,000°F in vacuum for 2 hr. Molybdenum heated at 2,150°F in H₂ for 2 hr and in vacuum at 4,000°F for 2 hr.

ing and slip casting in the fabrication of objects from refractory materials.

B. MICROELECTRONICS

In the manufacture of integrated circuits many of the active and passive components are built up in layers that are deposited sequentially from the vapor phase. Figure 15 illustrates the steps required to form a simple *p-n* junction on a *p*-type semiconductor substrate. In addition to the deposition of a mask and the incorporation of impurities from the vapor phase, practical devices may also require the deposition of other insulator, metal, and semiconductor layers. Much success in the formation of resistors, capacitors, and epitaxial semiconductor layers has been achieved with glow discharge deposition, either by sputtering or by chemical synthesis in the discharge.

In general, the sputtering of metals and oxide dielectrics has been reviewed previously (*10,35*). The technology in this field has expanded so widely in the last five years that only several recent developments in the glow discharge deposition of these materials, and in the formation of magnetic and superconducting thin films, are described here.

1. Dielectrics

In microelectronic devices, dielectric layers are used as masks to prevent the diffusion of impurities, as a means of electrically isolating components and protecting them from the atmosphere, and as thin-film insulators and capacitors. These materials are deposited as amorphous thin films. Silicon dioxide is the most widely used, but several inade-

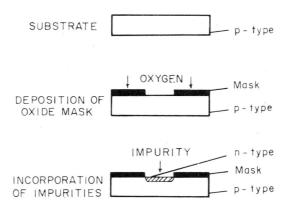

FIG. 15. Formation of a junction on a *p*-type substrate by sequential deposition from the vapor phase.

TABLE VII

METHODS OF PREPARATION OF SiO₂ THIN FILMS

Method	Temperature, °C	Oxide characteristics		Reference
		Refraction index[a]	Breakdown voltage[b]	
	Nondischarge:			
Steam oxidation	1200	1.466(5460 Å)		(38)
	1050	1.463(5890 Å)	8×10^6	(82)
Dry oxidation	1200			(38)
Hydrolysis of halides				
SiCl₄ + H₂ + CO₂ SiO₂ + 4 HCl	800–850	1.48(5460 Å)	2×10^7	(82)
SiCl₄ + 2 H₂ + 2 NO SiO₂ + N₂ + 4 HCl	950–1200	1.48–1.50(5460 Å)	$5\text{–}10 \times 10^6$	(83)
Decomposition of organooxysilanes				
Ethyl triethoxysilane, (CH₃CH₂)Si(OCH₂CH₃)₃	825	1.450(5460 Å)		(84)
Anodic oxidation: 0.04 N KNO₃-n-methylacetamide	25	1.362(5460 Å)	2.6×10^7	(88)
	glow discharge			
rf sputtering	100	1.476(5460 Å)		(85)
	450	1.473(5460 Å)		(85)
	450		1×10^7	(86)
Decomposition of organooxysilanes				
Tetraethoxysilane, (C₂H₅O)₄Si	75	1.55(460 Å)	$5\text{–}10 \times 10^6$	(87)
Plasma anodization	450	1.47(5460 Å)	1×10^7	(29,86)
Silane-nitric oxide reaction	25	1.49(5460 Å)		(43)

[a] Index of refraction (n) at wavelength indicated.
[b] Breakdown voltage (V/cm) of metal-oxide–metal capacitor.

quacies in it have led to the substitution of silicon nitride for some applications.

a. Silicon Dioxide

Table VII shows alternative methods that can be used to deposit thin films of SiO_2. The primary advantage of the glow discharge techniques is that the oxide is formed at a relatively low temperature. Heating of semiconductor devices can cause irreversible changes in their properties due to the redistribution of impurities. For example, the current-voltage characteristic of metal-insulator-semiconductor transistors is changed considerably by the diffusion of sodium ions during thermal treatment (36).

Of the glow discharge deposition techniques, sputtering is the best known. In contrast, plasma anodization has not yet been fully investigated, although several potential advantages have been found. Ligenza (29), using a microwave discharge, observed that oxide films on silicon could be grown up to 6000 Å thick with the substrate at about 300°C. The silicon substrate can be cleaned by bombarding it with positive ions prior to the oxidation. The oxide can serve as a mask as in Fig. 15, with the windows made by ion etching in the discharge (37). Thus the potential exists to deposit layers sequentially without opening the vacuum system to the atmosphere until a complete semiconductor device has been fabricated.

The physical and electrical properties of plasma-anodized silicon are similar to those of thermally grown oxides. However, Skelt and Howells (38) found a higher defect density in oxides produced by the former method. This may be due to surface features in the substrate that are reproduced in the oxide. Capacitors of aluminum oxide (39) and tantalum oxide (40) have been formed without pinholes, as evidenced by their high breakdown voltages and small leakage currents. One possible disadvantage of plasma anodization, as of thermal oxidation, is that some of the substrate is consumed. In silicon, the electrical properties in a region near the semiconductor-oxide interface may be changed.

Amorphous silica films can also be deposited by the decomposition reactions shown in Table VII, initiated by the energy supplied by the electrical discharge. For example, organosilanes are decomposed in an oxygen glow discharge to form amorphous films of SiO_2 (41). The substrate temperature is considerably lower than when thermal decomposition is used. However, the temperature must still be carefully controlled to prevent the incorporation of water vapor into the oxide and the polymerization of organic radicals on the substrate.

The strong effect of water vapor on the properties of SiO_2, and the ability of impurity ions to diffuse into the amorphous oxide, have recently provided the incentive to develop an alternative dielectric material. The compound selected is silicone nitride (Si_3N_4) and a great deal of effort has recently been devoted to developing methods of depositing it.

b. Silicon Nitride

Thin films of silicon nitride have a high dielectric strength, have high resistivity, and are chemically inert to most solvents. In addition, silicon nitride is an effective diffusion mask for the alkali metals that can penetrate silicon dioxide.

Silicon nitride cannot be formed by the thermal reaction between solid silicon and nitrogen gas, but amorphous thin films can be deposited on silicon by the pyrolytic decomposition of silane-ammonia mixtures (42). The substrate must be heated to 750°–1100°C. The reaction is

$$3\, SiH_4\, (g) + 4\, NH_3\, (g) = Si_3N_4\, (s) + 12\, H_2\, (g) \tag{4}$$

The solid can be deposited at much lower temperatures by carrying out reaction (4) in a glow discharge, or by reactive sputtering of a silicon cathode in a nitrogen discharge.

The glow discharge reaction between silane and ammonia can be brought about successfully even with a refrigerated substrate, but more desirable properties are obtained between 250°–300°C. Sterling and

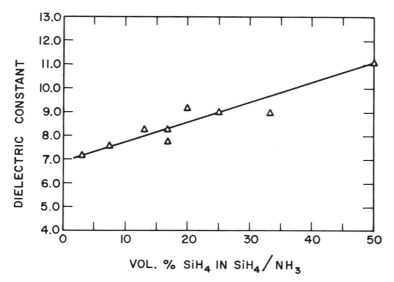

FIG. 16. Effect of SiH_4/NH_3 concentration on the dielectric constant of silicon nitride films produced in a glow discharge. Film thickness is 1000 Å (43).

Swann (43) found that silicon nitride films formed on a cold substrate are etched more rapidly in hydrofluoric acid than they are when the substrate temperature is 300°C. If the substrate is heated to 700°–900°C after deposition, the film resists acid etching even more. The dielectric constant of the silicon nitride also depends on the substrate temperature. It might be expected that changes in the silane-to-ammonia ratio would change the properties of the film by changing its stoichiometry. Figure 16 shows that the dielectric constant does change with the composition of the gas.

Similar results have been obtained with silicon nitride produced by reactive sputtering. Figure 17 shows that the power supplied to the discharge by the rf source affects the film properties.

All these facts suggest that the recombination of the species in the discharge to form Si_3N_4 occurs on the substrate. The relative concentration of reactants added to the discharge affects the stoichiometry of the condensed material. The power supplied by the discharge affects the energy of the species impinging on the surface, as well as their relative concentration in the gas. The substrate temperature influences the mobility of adsorbed species. If the mechanism of the glow discharge deposition involves recombination on the substrate followed by surface reaction, then the influence of the variables mentioned above on the structure and properties of the deposit can be explained qualitatively.

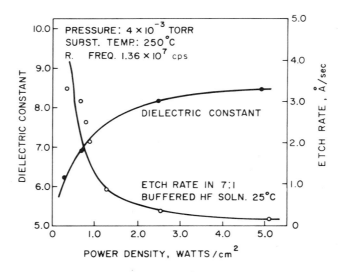

Fig. 17. Effect of rf power density on the dielectric constant and etch rate of reactively sputtered silicon nitride (44).

However, a detailed atomistic model to explain the experimental results has not yet been developed.

c. Other Materials

Among the other amorphous insulating films that can be produced by sputtering or by reactions in the glow discharge are aluminosilicate glasses, bismuth oxide, lead-tellurium oxide, zirconium oxide, aluminum oxide, vanadium oxide, and silicon carbide (45,46). The most important is tantalum oxide, which is produced successfully by the reactive sputtering of tantalum electrodes in inert gases doped with oxygen (47). The stoichiometry and the properties of the oxide are sensitive to the oxygen partial pressure.

Refractory metals such as tantalum produced by diode sputtering in an inert gas alone are suitable for thin film resistors because of their high resistivity and low temperature coefficient of resistivity. However, residual gases in the vacuum chamber such as O_2, N_2, CO, and CH_4 are a source of impurities that lead to the formation of different phases. These impurities affect the resistivity, but are not necessarily detrimental. In fact, nitrides formed with the tantalum can yield excellent resistive properties (11). Bias sputtering gives better control over the purity of tantalum resistors, but in the formation of oxide dielectrics deviations from stoichiometry cannot be avoided.

Improvement in thin-film insulators are still being sought by manufacturers of electronic components. Modifications of the deposition techniques and the development of new materials are both important.

2. Semiconductors

Silicon and germanium can be deposited epitaxially by sputtering or by evaporation. However, to deposit thin films without impurities that affect the electrical properties, sputtering is extremely useful (48). It is even more difficult to deposit compound semiconductors by evaporation, especially II-VI and III-V materials in which the vapor pressures of the components differ.

As an example of electrical discharge synthesis, aluminum nitride can be produced by sublimation of aluminum trichloride into a nitrogen glow discharge (49). With the use of diode sputtering, InSb, PbTe, Bi_2Te_3, CdTe, and other compounds can be deposited (50,51). Furthermore, amorphous thin films of solid solutions such as $Cd_xHg_{1-x}Te$ (51), $InAs_xSb_{1-x}$ (52), and PbS_xSe_{1-x} (53) can be produced. The latter are used as infrared radiation detectors. The materials can be crystallized by heating the substrate. Although Riggs (53) found that the stoichiometry of the deposit differed slightly from that of the starting material, and in general the film properties depend on the thickness and the substrate temperature, useful devices have been produced.

Other applications of electrical discharges in the production of semiconductor devices include the gas-phase doping of materials and the formation of transistor junctions by ion implantation. If silicon is deposited epitaxially by the glow discharge decomposition of silane or the reduction of silicon tetrachloride, then hydrides or chlorides, respectively, of the dopant can be added to the discharge. In this way boron or phosphorus can be incorporated as *p*- and *n*-type impurities (54). There is also the possibility of producing graded impurity profiles.

Alternatively, ion beams of boron or phosphorus can be used to implant impurities into silicon layers to form *p-n* junctions (55). An ion beam technique has also been developed for doping compound semiconductors, such as gallium arsenide (56) in the fabrication of photo diodes.

3. Superconductor and Magnetic Thin Films

The properties of superconductor and magnetic thin films are also greatly affected by impurities. Getter and bias sputtering (either dc or asymmetric ac) are therefore very useful for these materials. To produce niobium, tantalum, and vanadium thin films with superconducting transition temperatures near or above those of the bulk solid, O_2, N_2, and H_2O must be eliminated (57). The results in Table VIII show that ultraclean vacuum conditions are not required if the proper sputtering techniques are used.

The primary reasons for sputtering magnetic materials are that alloys can be deposited without changes in composition and uniform coverage

TABLE VIII

DEPOSITION OF NIOBIUM THIN FILMS (57)

Deposition technique	Film thickness, Å	Resistivity, ohms		
		300°K	77°K	4.2°K
Evaporated niobium	No data	No data	No data	Nonsuperconducting
Conventional dc sputtering	7,500	No data	No data	Nonsuperconducting
Bias sputtering at 60 V, dc	1,700	17	19	20
Bias sputtering at 60 V, dc	1,700	26	28	29
Bias sputtering at 200 V, dc	1,700	6.2	5.6	Superconducting
Bias sputtering at 200 V, dc	1,700	5.0	4.8	Superconducting
Bias sputtering at 80 V, dc	1,700	2.6	2.3	Superconducting
Bias sputtering at 100 V, dc	1,700	1.7	1.3	Superconducting
With titanium getter in operation	900	6.5	5.2	Superconducting

of the substrate can be obtained. These factors are important in the preparation of ferromagnetic films, such as permalloys, used for information storage in computers (50,58).

Ferrites can be sputtered directly (59), or the alloys can first be deposited by sputtering and then thermally oxidized (60). Naoe and Yamanaka (60) have shown the versatility of the direct sputtering in the formation of ferrites containing Ni, Zn, and Fe.

C. EXTRACTIVE METALLURGY

Electrical discharges can be used to carry out pyrometallurgical reactions. At the operating temperature of an arc plasma jet, refractory metal ores can be decomposed or reduced to produce the metal directly, or simple compounds from which the metal can be extracted. In contrast to spray coating, in this application it is usually desirable to completely vaporize the reactants, since homogeneous gas phase reactions are rapid and produce simple products from even the most complex ores.

The ore can be mixed with sufficient carbon to make it a consumable anode, but it is more economical to feed the ore continuously into the plasma (21). There is no contamination from products of combustion, slag, or the brick linings of conventional furnaces.

Although relatively simple chemical species exist in the vapor phase, the high-temperature capability of the plasma cannot be used effectively if back reactions occur upon cooling. The maintenance of a metastable equilibrium in the reaction products has been the major difficulty in the development of plasma techniques in extractive metallurgy. Examples of the types of reactions that are sought in the plasma reduction or decomposition of ores are shown in Table IX.

1. Decomposition of Ores

Several ores have been successfully decomposed in the high intensity arc. Rhodonite, a manganese silicate, condenses from the plasma temperature to form MnO and SiO_2. Since the compounds are quenched from the vapor phase, the particles are very fine, and the manganous oxide dissolves rapidly in cold hydrochloric acid. By leaching the product mixture, 95% recovery of the manganese can be obtained (61). Similarly, beryl, a silicate of beryllium and aluminum, is decomposed to BeO, Al_2O_3, and SiO_2 (62). Molybdenum sponge can be formed by the dissociation of molybdenite, MoS_2.

2. Reduction of Ores

Direct reduction of the oxides of Al, Mg, Be, B, Ti, Zr, and several other metals, which cannot be done in conventional smelters, can be

TABLE IX

ORE TREATMENT IN PLASMA JETS

Process and typical ore	Reaction[a]	Metals	Comment
Thermal decomposition			
Rhodonite	$MnSiO_3 \rightarrow MnO + SiO_2$	Mn, U, Be, Li, Zr	The simple oxides are leached to recover the metal
Molybdenite	$MoS_2 \rightarrow Mo + S_2$		
Direct reduction			
Dolomite	$MgCO_3 + 2\,C \rightarrow Mg + 3\,CO$	Al, Mg, Be, B, Nb, Si, Mn, Ti, Zr	Ores can be vaporized in the torch or the metal separated as a liquid
Alumina	$Al_2O_3 + C + H_2 \rightarrow Al + H_2O + CO$		
Zirconia	$ZrO_2 + 2\,H_2 \rightarrow Zr + 2\,H_2O$		
Vapor phase halogenation			
Alumina	$Al_2O_3 + 3\,C + 6\,HCl \rightarrow 2\,AlCl_3 + 3\,CO + 3\,H_2$	Be, Fe, Al, Nb, Si, Ta	The metal halides are recovered by fractional condensation
Bromellite	$BeO + C + Cl_2 \rightarrow BeCl_2 + CO$		

[a] These reactions indicate what can be attained in the plasma. Practical recoveries and commercial utilization in many cases have not been attained. See text.

done in the plasma. Elemental vapors and carbon monoxide are formed at the plasma temperature from mixtures of the oxide ore and carbon. Mass spectrometer studies of MgO reveal only Mg, O, C, and CO in the vapor phase (62). However, the metal oxide is formed again upon cooling. This is also the primary limitation in the direct reduction of alumina and zirconia in a hydrogen plasma. Owing to the rapid recombination, only a small yield of metal is obtained, even though complete reduction takes place in the vapor phase.

An alternative to the direct reduction in the vapor phase is to keep the temperature below the boiling point of the metal, and to collect the metal as a molten liquid (63). This procedure has been used successfully in the production of columbium.

The separation can often be improved if the reduction is done with a halogen gas added to the plasma. The oxides are converted to halides, and the latter condense upon cooling. The products of the vapor phase halogenation can be separated readily by fractional condensation of the metal halides. In niobium oxide, for example, the equilibrium favors the formation of $NbCl_5$ even after the products are cooled (64). As is discussed in greater detail in Section III,D, metal chlorides such as BCl_3 and $AlCl_3$ can be used to deposit metals of high purity. A glow discharge is satisfactory for the reactions since the temperatures are relatively low.

The utilization of electrical discharges in extractive metallurgy is not yet successful to the point where pure metal products can be produced on a commercial scale. Improvement in the separation of the products during quenching, and in the efficiency of conversion of electrical power to heat would make this application extremely useful. The potential exists to miniaturize extractive metallurgical processes.

D. CHEMICAL SYNTHESIS

1. Metals and Inorganic Compounds

The techniques for producing metal chlorides in electrical discharges have already been described. Reduction of the halide in an electrodeless glow discharge or in a plasma torch can be used to deposit many metals of high purity. Similarly, metal compounds can be produced by reacting the vaporized halides with gases such as ammonia and nitrogen to give metal nitrides, or with oxygen to give metal oxides. A system for depositing boron nitride in an rf electrodeless discharge, which could also be used for metals and other compounds, is illustrated in Fig. 12.

If a powdered metal is fed into a plasma torch with nitrogen or methane, metal nitrides or carbides can be produced. Yields approaching 100% are obtained for TiN, and lower yields for WN, WC, Mg_3N, and

TaC (65). Boron carbide has been prepared by using a mixture of propane and butane as the source of carbon (66).

According to McTaggart (15) chlorides of the Group I metals Li, Na, K, and Cs dissociate in electrical discharges generated in either inert gases or hydrogen. Recovery of the metals even though the halogen gases are ordinarily very reactive is said to be possible because the gas exists primarily as unreactive negative ions in the region of the discharge where the metal is deposited. Deposits of Be, Mg, Ca, Sr, and Ba are also formed when their chlorides are vaporized into the discharge. To obtain appreciable metal yields from BCl_3, $AlCl_3$, and the rare earth chlorides $CeCl_3$ and $LaCl_3$, hydrogen is needed as the reducing agent in the discharge. The glow discharge reaction between a gaseous metal compound and hydrogen can also be used to produce Si, Ge, As, Sb, Te, Pb, Bi, Nb, Ti, Ta, Mo, and W. It is likely that the technique is even more versatile.

Boron can be produced with purity up to 99.9% (67). The techniques of producing boron in both a glow discharge (67) and a corona discharge (68) by the hydrogen reduction of boron trichloride and boron tribromide have recently been applied to the making of boron filaments to increase the strength of composite structures. The feasibility of depositing boron on a continuously moving substrate has been demonstrated. On a tungsten wire substrate, the reaction is reported by Wales (68) to be electrochemical, positive boron ions being reduced on the negatively charged substrate. However, many variables affect the deposition. Among these are the reactor design, gas flow rates, impurities in the gases, substrate surface, and electrical power input. Therefore, a method to control the process carefully is required if this technique is to be scaled up for commercial production of coated filaments.

2. Polymerization

A low pressure dc or electrodeless discharge can be used to initiate the polymerization of many organic vapors, and adherent polymer coatings can be deposited on almost any substrate. For example, fabric, metal, and paper strips, or single fibers of metals and ceramics can be coated continuously up to thicknesses of several microns. Applications also exist in microelectronics, where polymers are used as thin-film capacitors and for encapsulation of semiconductor devices.

Although many monomers have been studied, the best results thus far are for the polymerization of styrene ($CH_2{=}CH{-}C_6H_6$). Films that are free from pinholes have been produced with thicknesses as low as 100 Å. Goodman (69) attributes this to the fact that the electric field is stronger in the thin areas, leading to an increased rate of deposition that mends the polymer film in these areas. It is likely that ionized molecules and excited species are adsorbed on the substrate surface and polymerize the monomer that is also adsorbed. This is based on evidence presented by Williams and Hayes (16) that the rate of polymerization is related to the amount of monomer adsorbed, and is independent of the pressure

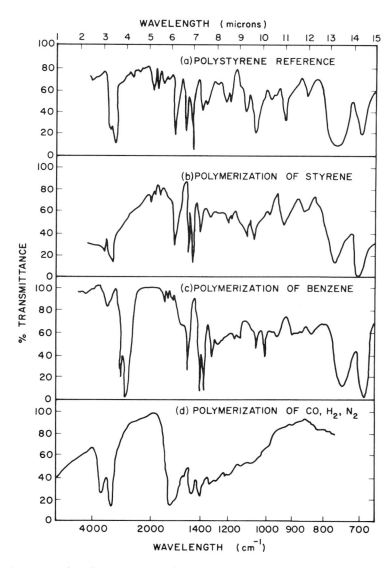

FIG. 18. Infrared spectra of polymers produced by glow discharge synthesis: (a,b), Williams and Hayes (16); (c), Neiswender (70); (d), Hollahan and McKeever (72).

above several Torr. Polystyrene films are formed with a high degree of crosslinking, which is probably the result of electron bombardment and ultraviolet irradiation of the film.

The reaction of vaporized benzene in a helium rf discharge also produces polystyrene (70). This is confirmed by the infrared spectra shown in Fig. 18. It is an expected result, since the discharge can initiate reactions that convert the benzene to styrene, with acetylene as the intermediate:

$$
\text{(benzene)} \rightleftharpoons 3\,HC{\equiv}CH \xrightarrow{HC{\equiv}HC} \text{(styrene: } CH{=}CH_2 \text{)} \tag{5}
$$

The polymerization then occurs by the mechanism described above for the styrene monomer.

Glow discharge polymerization of hydrocarbon-hydrogen mixtures can be used to deposit solid polymers with varying hydrogen-to-carbon ratios by adjusting the ratio of the starting materials. Vastola and Greco (12) found that the polymer contains the same ratio as the reactants for $H/C < 1.6$. This same effect can be used in depositing organo-silicon polymers. At one extreme tetraethoxysilane can be decomposed to silicon dioxide in an rf discharge, as discussed in Section II,B. Other siloxanes are known to polymerize, and by adding these to the discharge, polymer layers can be gradually built up. The potential, therefore, exists to deposit polymer coatings with varying hydrocarbon to silicon ratios. Coatings of this kind can be used to increase the adhesion between a glass fiber and a polymer matrix (71).

The glow discharge reactions considered thus far use hydrocarbon reactants. An interesting possibility that has been investigated by Hollahan and McKeever (72) is synthesis from inexpensive starting materials such as carbon monoxide, hydrogen, and nitrogen. Although the gas flow rates and the discharge pressure must be carefully controlled, and water vapor must be eliminated, polymerization does occur. The infrared transmission spectrum of the polymer is also shown in Fig. 18.

E. CRYSTAL GROWTH

The use of the plasma-torch and the hollow-cathode-zone refining methods has broadened the possibilities for producing synthetic materials and has permitted the physical properties of refractory materials to be studied as single crystals.

Sapphire, zirconia, and niobium were grown by Reed (24) in a plasma furnace similar to the one shown in Fig. 10. The rate of withdrawal of the crystal, gas flow rate, powder feed rate, and particle size must be carefully controlled to avoid deposition of polycrystalline material. Reed reported the growth of crystals up to 15 mm in diameter and 90 mm long at a rate of 20–50 mm/hr. Ruby and sapphire crystals grown by this method have dislocation densities similar to crystals grown by the Verneuil technique in an oxy-hydrogen flame. However, materials such as W, Mo, Cb, TiC, and TiB$_2$ that have been grown in the plasma torch cannot be produced by flame fusion (7).

Zone refining by the hollow-cathode floating-zone method has been proposed as an alternative to growing single crystals of high-melting-point materials (73,74). Success has been obtained with sapphire, where dislocation densities of $10^3/cm^3$ are reported. This is two orders of magnitude less than for crystals grown by the Verneuil technique. Although the hollow cathode is similar to the electron beam used in zone refining of some refractory materials, compounds such as yttria (Y_2O_3) cannot be grown by ordinary electron beam techniques since their vapor pressures at the elevated temperatures required are above the 10^{-4} Torr used in electron beam zone refining. The hollow cathode, however, can operate in an oxygen atmosphere at several Torr. Oxides of Er, Th, La, Ti, and Zr have also been grown as single crystals (74).

F. Production of Powders

Submicron-size particles of metals, carbides, nitrides, oxides, intermetallic compounds, and alloys can be produced by vaporizing material into an arc plasma torch and then rapidly quenching in a cooled gas or liquid. Fine particles of carbon have been produced by passing hydrocarbons through an argon plasma jet (32). Similarly, nickel powder, or mixtures of nickel and nickel oxide, can be produced in the size range 30–300 Å by decomposing nickel carbonyl ($Ni(CO)_4$) in the plasma (75). By the direct reaction of the elements, titanium nitride, with particle sizes between 0.75 and 7.5 microns, has been produced with 100% yield.

Nucleation from the vapor phase occurs very rapidly in the plasma, producing small-diameter particles. Spheroidization can also be accomplished in the plasma torch, and has been applied to the production of nuclear fuels (76).

At temperatures below the boiling point, melting and resolidification produce particles of larger sizes. Although the size distribution is difficult to control, Moss and Young (32) show pictures of Mo spheres in sizes between 6 and 8 μ. By allowing molten metal to fall into water or lubricating oil, powders of Cr, Mo, and W have each been produced in the range 100–450 μ (77). The Cr and Mo were spherical but the W was not. As in other applications of high temperature plasmas the cooling conditions must be controlled to obtain reproducible results. It can also be expected that for many compounds stoichiometric differences will exist between rapidly quenched powder and the bulk feed material.

G. Sintering

The use of high-energy sparks for sintering has been developed within the last few years (78). The process is similar in principle to the electrical discharge machining described in Section I. Spark sintering of powder placed between graphite electrodes shaped to conform to the finished object is reported to be extremely effective for materials such as aluminum, which has a natural oxide film on its surface (79). During the transient spark discharge the particles are fused and coalesce with densities between 65% and 99% of theoretical. Heat and static pressure are applied simultaneously.

This technique has permitted the direct bonding of sintered metal particles to a metallic substrate, and has been used to bond WC powder to tool steels. Proposed applications include the spark sintering of cast iron powders into piston rings, and the bonding of sintered diamond and WC powders directly to grinding wheels.

H. Other Applications

Although extensive results have not yet been obtained, electrical discharges can be used to chemically purify metals and alloys and to modify their surface properties. For example, sparks generated as in electrical discharge machining lead to surface hardening in both ferrous and nonferrous metals (4,80), due to melting and resolidification on the surface. Phase transformations in the surface region can also occur. Nitriding and carburization can be carried out with an aqueous ammonia solution or a hydrocarbon, respectively, between the discharge electrodes. Harder and more wear-resistant surface layers are produced by nitriding in a nitrogen glow discharge, or carburization in a discharge of a hydrocarbon gas. By this technique it is possible to modify the properties of the inner wall of cylindrical tubing.

Surfaces can be cleaned in glow- or arc-discharges by chemical reaction between the surface and hydrogen, or by sputtering in an inert gas. Improvements in the purity of materials can also be obtained by using a plasma arc torch to melt and refine refractory metals. Volatile impurities are removed preferentially. For example, the oxygen content of steels can be reduced by several orders of magnitude.

These applications follow almost directly from the general usefulness of electrical discharges to produce high temperatures and to create chemically reactive species.

IV. Prospects for Future Developments

The versatility of the electrical discharge as a heat source for melting and evaporation and as a medium for carrying out chemical reactions suggests that new applications will be developed. The 1960 report of the Materials Advisory Board of the National Academy of Sciences recommended that greater emphasis should be given to an understanding of the basic phenomena that occur in plasmas in order to acquire the insight needed to develop practical applications (7). Since this report was published new fundamental and practical achievements have occurred. At present the major need is to eliminate the shortcomings in applications that have already been realized.

The application of thermal plasmas in extractive metallurgy has considerable potential, but if this potential is to be developed, the efficiency of product recovery must be increased by reducing the recombination that occurs during cooling of the elemental constituents in the high-temperature plasma. A more fundamental understanding of the kinetics of the reactions that take place during quenching from the plasma temperature is required for the solution to this problem.

The use of the plasma torch broadens the possibilities for the development of coatings and single crystals of synthetic materials. Plasma-sprayed coatings on inexpensive base materials appear to be economically attractive. Graded coatings that increase the compatibility of materials, and polymers that can be sprayed without subsequent thermal curing will probably increase in importance. In addition, as exotic fuels and high-temperature engines are developed, new materials of construction will be needed. It is therefore likely that in aerospoce applications plasma coating and fabrication techniques for producing heat and corrosion resistant materials will become even more common. However, alternative methods of fabrication will probably be improved also. For example, electrochemical machining, ultrasonic machining, and powder metallurgy already compete with electrical discharge machining. The choice of any technique depends on the particular application, so that a choice of fabrication methods must be available as demands for new materials originate. The availability of powders produced by electrical discharges suggests an increase in the future use of powdered refractory materials.

In fact, the plasma spray forming of rocket and missile parts has already increased because starting materials are available.

Thermal plasmas may find additional uses in purification, melting, and casting. Impurities such as phosphorus, sulfur, and arsenic are deleterious to the mechanical properties of newly developed high-strength steels suitable for use in oceanographic studies at great depth (81). However, these impurities are difficult to remove. Attempts to obtain very high purity may be significantly aided by the ability of the plasma torch to volatilize and dissociate impurities, and to fractionally distill solids.

The major innovations in applying electrical discharges to the production of materials will probably be in the area of chemical synthesis. The approach of Hollahan (72) to the production of organic polymers, using carbon monoxide, hydrogen, and nitrogen as the starting materials, is an outstanding example of this future potential. Again, a more fundamental understanding of the reaction processes that occur in the discharge is required before full utilization of this potential can be made. If reactions occurring in electrical discharges can be made to yield thermodynamically unstable products, materials may be produced with properties unequaled by conventional materials. The large driving force in reactions between the plasma and condensed phases, as well as in reactions in the gas phase, may produce these useful new products. Less exotic materials, but ones that are none the less important for new applications, can be produced in electrical discharges. For example, continuously coated fibers for filament-reinforced composites have considerable potential. One general problem, in addition to controlling the properties of the materials produced, is the scale-up of electrical discharge reaction processes to commercial operations. Here, more fundamental knowledge of electrical discharge chemistry is vital.

In the microelectronics industry the continuing increase in integrated circuit applications will call for improved deposition techniques for thin-film semiconductors, dielectrics, and magnetic materials. Also, new electronic materials such as high-transition-temperature superconductors may well be produced with controlled purity by modifications of existing sputtering methods. Similarly, resistors with higher specific resistivity and lower temperature coefficient of resistivity, capacitors with higher breakdown voltages and lower leakage currents, and diffusion barriers with lower permeabilities may be made by improved glow discharge deposition techniques. There is already a great deal of research directed toward an understanding of the relationship between the sputtering conditions and the properties of the products. In addition to sputtering and chemical synthesis in glow discharges, the use of plasma spraying to deposit films for integrated circuits (*81a*) illustrates the very wide range of electrical discharge applications in producing electronic materials.

As the ozonizer was an early application of electrical discharges in the purification of water, reactions in electrical discharges may some day be successful in removing pollutants from the air, or in purifying exhaust gases before they are introduced into the atmosphere. Now that most of the equipment problems in generating discharges have been overcome, future applications in the production of materials depend primarily on the ingenuity and imagination of scientists and engineers to develop a complete understanding of electrical discharge phenomena and to translate this knowledge into the solution of practical problems.

REFERENCES

1. L. B. Loeb, "Fundamental Processes of Electrical Discharges," p. 543. Wiley, New York, 1959.
2. F. Park, Nontraditional machining. *Intern. Sci. Technol.* 11, 22 (1963).
3. W. F. Courtis, Electrical discharge metal forming. *Mech. Eng.* 84, No. 10, 46 (1962).
4. N. Yutani and A. L. Pickrell, Electrical discharge machining of difficult alloys. *Metal Progr.* 82, No. 2, (1962).
5. A. M. Howaston, "An Introduction to Gas Discharges." Pergamon, Oxford, 1965.
6. M. L. Thorpe, Plasma jet: progress report II. *Res./Develop.* 12, No. 6, 77–89 (1961).
7. Materials Advisory Board, National Academy of Sciences. "Development and Possible Applications of Plasma and Related High Temperature Generating Devices," Rept. MAB-167-M. Washington, D. C., 1960.
8. G. K. Wehner, Sputtering by ion bombardment. *Advan. Electron. Electron Phys.* 7, 239–298 (1955). (L. Marton, ed.), Academic Press, New York, 1955.
9. S. P. Wolsky, Sputtering mechanisms. *In* "Transactions of the 10th Vacuum Symposium, American Vacuum Society" (G. H. Bancroft, ed.), pp. 309–315. Macmillan, New York, 1963.
10. L. Holland, "Vacuum Deposition of Thin Films." Wiley, New York, 1958.
11. N. Schwartz, Reactive sputtering. *In* "Transactions of the 10th Vacuum Symposium, American Vacuum Society" (G. H. Bancroft, ed.), pp. 325–334. Macmillan, New York, 1963.
12. F. J. Vastola and B. Greco, Hydrocarbon reactions in a plasma jet. *In* "Symposium on Chemical Phenomena in Gases," Abstract No. 10I, 147th Meeting, Am. Chem. Soc., Philadelphia, 1964.
13. R. F. Baddour and R. S. Timmens, eds., "Applications of Plasmas in Chemical Processing." MIT Press, Cambridge, Massachusetts, 1967.
14. F. K. McTaggart, "Plasma Chemistry in Electrical Discharges." Elsevier, Amsterdam, 1967.

15. F. K. McTaggart, The dissociation of metal halides in electrical discharges. *In* "Symposium on Chemical Reactions in Electrical Discharges" (H. R. Linden and I. Wender, eds.), pp. 156–160. Am. Chem. Soc., Miami, 1967.
16. T. Williams and M. W. Hayes, Polymerization in a glow discharge. *Nature* **209**, 769–773 (1966).
17. P. L. Spedding, Chemical synthesis by gas phase discharge. *Nature* **214**, 124–126 (1967).
18. A. D. MacDonald, "Microwave Breakdown in Gases." Wiley, New York, 1966.
19. H. Puschner, "Heating With Microwaves." Springer, New York, 1966.
20. N. N. Ault and W. M. Wheildon, Modern flame-sprayed ceramic coatings. *Modern Materials* **2**, 63–106 (1960).
21. S. Korman, High-intensity arcs. *Intern. Sci. Technol.* **6**, 90–98 (1964).
22. L. W. Davis, How to deposit metallic and nonmetallic coatings with the arc plasma torch. *Metal Progr.* **83**, No. 3, 107 (1963).
23. Thermal Dynamics Corp., Lebanon, New Hampshire. Bulletin 155A.
24. T. B. Reed, Growth of refractory crystals using the induction plasma torch. *J. Appl. Phys.* **32**, 2534 (1961).
25. T. B. Reed, Plasma torches. *Intern. Sci. Technol.* **6**, 42 (1962).
26. C. F. Powell, J. H. Oxley, and J. M. Blocher, Jr., eds., "Vapor Deposition." Wiley, New York, 1966.
27. L. I. Maissel and P. M. Schaible, Thin films deposited by bias sputtering. *J. Appl. Phys.* **36**, 237–242 (1965).
28. R. Frerichs, Superconductive films made by protective sputtering of tantalum and niobium. *J. Appl. Phys.* **33**, 1898 (1962).
29. J. R. Ligenza, Silicon oxidation in an oxygen plasma excited by microwaves. *J. Appl. Phys.* **36**, 2703–2707 (1965).
30. R. H. Heil, Jr., Advances in zone refining. *Solid State Technol.* **11**, No. 1, 21–28 (1968).
31. C. A. Krier and W. A. Baginski, Coated refractory metals. *Modern Materials* **5**, 1–94 (1965).
32. A. R. Moss and W. J. Young, The role of arc-plasma in metallurgy. *Powder Met.* **7**, 261–289 (1964).
33. M. A. Levinstein, A. Eisenlohr, and B. E. Kramer, Properties of plasma-sprayed materials. *Welding J.* (*Welding Res. Suppl.*) **40**, 8s–13s (1961).
34. W. A. Spitzig and G. W. Form, Effects of sintering on the physical and mechanical properties of arc plasma-sprayed tungsten. *Trans. Met. Soc. AIME.* **230**, 67–70 (1964).
35. L. I. Maissel, Sputtering. *Thin Films* **3**, (1966).
36. R. C. G. Swann, R. R. Mehta, and T. P. Cauge, The preparation and properties of thin film silicon-nitrogen compounds produced by radio frequency glow discharge reactions. *J. Electrochem. Soc.* **114**, 713–717 (1967).
37. R. T. C. Tsui, Ion etch technique and its applications. *Solid State Technol.* **10**, No. 12, 33–38 (1967).
38. E. R. Skelt and G. M. Howells, The properties of plasma-grown SiO_2 films. *Surface Sci.* **7**, 490–495 (1967).
39. G. J. Tibol and R. W. Hull, Plasma anodized aluminum oxide films. *J. Electrochem. Soc.* **111**, 1368–1372 (1964).
40. T. A. Jennings and W. McNeill, Gas phase anodization of tantalum. *J. Electrochem. Soc.* **114**, 1134–1137 (1967).

41. D. R. Secrist and J. D. Mackenzie, Deposition of silica films by the glow discharge technique. *J. Electrochem. Soc.* 113, 914–920 (1966).
42. V. Y. Doo, D. R. Kerr, and D. Nichols, Property changes in pyrolytic silicon nitride with reactant composition changes. *J. Electrochem. Soc.* 115, 61–64 (1968).
43. H. F. Sterling and R. C. G. Swann, Chemical vapor deposition promoted by r.f. discharge. *Solid State Electron.* 8, 653–654 (1965).
44. S. M. Hu and L. V. Gregor, Silicon nitride films by reactive sputtering. *J. Electrochem. Soc.* 114, 826–833 (1967).
45. M. L. Lieberman and R. C. Medrud, Reactively sputtered oxide films. *J. Electrochem. Soc.* 116, 242–247 (1969).
46. T. L. Chu, Chemical deposition of dielectrics for thin film circuits and components. *Solid State Technol.* 10, No. 5, 36–41 (1967).
47. F. Vratny, Deposition of tantalum and tantalum oxide by superimposed rf and d-c sputtering. *J. Electrochem. Soc.* 114, 505–507 (1967).
48. K. E. Haq, Deposition of germanium films by sputtering. *J. Electrochem. Soc.* 112, 500–502 (1965).
49. G. A. Cox *et al.*, On the preparation, optical properties, and electrical behavior of aluminum nitride. *J. Phys. Chem. Solids.* 28, 543–548 (1967).
50. M. H. Francombe, Preparation and properties of sputtered films. *In* "Transactions of the 10th Vacuum Symposium, American Vacuum Society" (G. H. Bancroft, ed.), pp. 316–324. Macmillan, New York, 1963.
51. H. Kraus, S. G. Parker, and J. P. Smith, $Cd_xHg_{1-x}Te$ films by cathodic sputtering. *J. Electrochem. Soc.* 114, 616–619 (1967).
52. R. F. Potter and G. G. Kretschmar, Optical properties of $InAs_ySb_{1-y}$ layers. *Infrared Phys.* 4, 57–65 (1964).
53. B. A. Riggs, Vacuum-deposited thin films of the type PbS_xSe_{1-x}. *J. Electrochem. Soc.* 114, 708 (1967).
54. J. H. Alexander and H. F. Sterling, Semiconductor epitaxy-gas phase doping by electric discharge. *Solid State Electron.* 10, 485–490 (1967).
55. K. E. Manchester, S. B. Sibley, and G. Alton, Doping silicon by ion implantation. *Nucl. Inst. Methods* 38, 169 (1965).
56. P. E. Roughan and K. E. Manchester, Properties of ion-implanted GaAs diodes. *J. Electrochem. Soc.* 116, 278–279 (1969).
57. J. M. Seeman, Bias sputtering: its techniques and applications. *Vacuum* 17, No. 3, 129–137 (1966).
58. E. Kay, Uniaxial anisotropy permalloy films grown in a glow discharge on sputtered tantalum. *Nature* 202, 788–789 (1964).
59. W. D. Westwood *et al.*, Single crystal ferrite films prepared by cathode sputtering. *J. Am. Ceram. Soc.* 50, 119–123 (1967).
60. M. Naoe and S. Yamanaka, Magnetic properties of ferrite films deposited by vacuum-arc discharge. *Jap. J. Appl. Phys.* 6, No. 8, 1029–1031 (1967).
61. V. Harris *et al.*, Arc decomposition of rhodonite. *J. Electrochem. Soc.* 106, 874 (1959).
62. C. P. Beguin, A. S. Kanaan, and J. L. Margrave, Plasma chemistry. *Endeavor* 23, No. 89, 55–60 (1964).
63. C. Sheer, S. Korman, and J. O. Gibson, Process for reduction of ores to metals, alloys, interstitial and intermetallic compounds. U. S. Patent 3,101,308 (August 1963).

64. I. H. Warren and H. Shimizu, Applications of plasma technology in extractive metallurgy. *Can. Mining Met. Bull.* **58**, 551–560 (1965).
65. C. S. Stokes, Chemistry in high temperature plasmas. *In* "Symposium on Chemical Reactions in Electrical Discharges" (H. R. Linden and I. Wender, eds.), pp. 312–331. Am. Chem. Soc., Miami, 1967.
66. F. M. Bosch, Synthesis of boron carbide and nitride. *Silicates Ind.* **27**, 587–590 (1962).
67. A. E. Hultquist and M. E. Sibert, Glow discharge deposition of boron. *In* "Symposium on Chemical Reactions in Electrical Discharges" (H. R. Linden and I. Wender, eds.), pp. 161–178. Am. Chem. Soc., Miami, 1967.
68. R. D. Wales, Plating in a corona discharge. *In* "Symposium on Chemical Reactions in Electrical Discharges" (H. R. Linden and I. Wender, eds.), pp. 161–178, pp. 179–902, Am. Chem. Soc., Miami, 1967.
69. J. Goodman, The formation of thin polymer films in the gas discharge. *J. Polymer Sci.* **44**, 551–552 (1960).
70. D. D. Neiswender, The polymerization of benzene in a radio-frequency discharge. *In* "Symposium on Chemical Reactions in Electrical Discharges" (H. R. Linden and I. Wender, eds.), pp. 274–281. Am. Chem. Soc., Miami, 1967.
71. Personal communication. Commonwealth Scientific Corp., Alexandria, Virginia, 1968.
72. J. R. Hollahan and R. P. McKeever, Radiofrequency electrodeless synthesis of polymers: reaction of Co, N_2, and H_2. *In* "Symposium on Chemical Reactions in Electrical Discharges" (H. R. Linden and I. Wender, eds.), pp. 254–258, Am. Chem. Soc., Miami, 1967.
73. W. Class, H. B. Nesor, and G. T. Murray, Preparation of oxide crystals by a plasma float-zone technique. *J. Phys. Chem. Solids. Suppl.* **1**, 75–80 (1967).
74. W. Class, Hollow cathode float-zone refining. *Res./Develop.* **18**, No. 9, 56–60 (1967).
75. T. B. Selover, Jr., Properties of nickel fume generated in a plasma jet. *AIChE J.* **10**, 79–82 (1964).
76. J. O. Gibson and R. Weidman, Chemical synthesis via the high intensity arc process. *Chem. Eng. Progr.* **59**, No. 9, 53–56 (1963).
77. A. N. Krasnov, G. V. Samsonov, and V. M. Sleptosov, Preparation of powder of copper, molybdenum, and tungsten by atomization in a plasma jet. *Izv. Akad. Nauk. SSSR, Metally. 1965* **3**, 70–72.
78. K. Inoue, "Electrical Discharge Sintering." U. S. Patent 3,241,956 (March, 1966); "Apparatus For Electrically Sintering Discrete Bodies," U. S. Patent 3,250,892 (May, 1966).
79. G. DeGroat, What's new in powder metallurgy. *Am. Machin.* **10**, No. 22, 151–152 (1966).
80. N. C. Welsch, Surface hardening of nonferrous metals by spark discharge. *Nature* **181**, 1005 (1958).
81. P. B. Lederman and D. H. Hallas, Materials: key to exploiting the oceans. *Chem. Eng.* **75**, No. 12, 107 (1968).
81a. D. H. Harris and R. J. Janowiecki, Arc-plasma deposits may yield some big microwave dividends. *Electronics* **43**, No. 3, 108–115 (1970).
82. M. J. Rand and J. L. Ashworth, Deposition of silica films by the carbon dioxide process. *J. Electrochem. Soc.* **113**, 48–50 (1966).
83. M. J. Rand, A nitric oxide process for the deposition of silica films. *J. Electrochem. Soc.* **114**, 274–277 (1967).

84. J. Klerer, On the mechanism of the decomposition of silica by pyrolytic decomposition of silanes. *J. Electrochem. Soc.* **112**, 502–506 (1965).
85. P. D. Davidse, Theory and practice of r.f. sputtering. *Vacuum* **17**, No. 3, 139–145 (1967).
86. R. W. Wilson, Thin film passive elements for monolithic integrated circuits. *Solid State Technol.* **10**, 21–26 (1967).
87. S. W. Ing and W. Davern, Glow discharge formation of silicon oxide and the deposition of silicone oxide thin film capacitors by glow discharge techniques. *J. Electrochem. Soc.* **112**, 284–288 (1965).
88. L. Young, "Anodic Oxide Films," p. 332. Academic Press, New York, 1961.

PYROLYTIC GRAPHITE

William H. Smith and Donald H. Leeds

Super-Temp Company, Santa Fe Springs, California

I. Introduction

Rapid progress has been made in the chemical vapor deposition (CVD) process during the past twenty years (1). Among outstanding contributions has been the commercial production of carbon by the decomposition of a hydrocarbon gas on a hot surface. The CVD material so produced is generally referred to as "pyrolytic graphite." Although the material is not a true graphite in the crystallographic sense, the term is used extensively and is now generally accepted. Pyrolytic graphite is unique among high-temperature materials in that its properties are extremely anisotropic; i.e., its properties vary depending on the direction in which they are measured.

Unlike many of our "newer" materials, pyrolytic graphite has a very long history. It was first produced in the late 1800's by Voelker (2), Sawyer (3), and others for lamp filaments. It was not until the 1950's, however, as a result of the need for improved nose cones, rocket nozzles, and nuclear reactors, that pyrolytic graphite was produced in massive shapes.

Well over 200 technical papers have appeared since 1950 on the various properties of pyrolytic graphite; however, little information has been published on the manufacturing process since most producers consider their techniques to be proprietary. Graphite in general has been covered in Volume 4 of *Modern Materials* (4).

The following presentation deals with many of the manufacturing variables, properties, and uses of pyrolytic graphite. Its structure and its effect on properties are also considered. Finally, the latest data are presented on the physical, thermal, electrical, and chemical properties, appropriate methods of quality control, and the most frequent applications in the aerospace and commercial markets.

As mentioned above, Sawyer and others around 1880 produced filaments of pyrolytic graphite by the CVD process. The filaments were extensively used in electric lamps until the early 1900's when methods were developed for the production of lamp filaments of tungsten. Because of its lower cost, greater toughness, and longer life, tungsten replaced carbon in the filament lamp. Some further use of thin films of pyrolytic graphite was made by the electrical industry in thin film resistors. Much of our early knowledge of the deposition process was a result of publications in this field.

Grisdale and Pfister (5) of the Bell Laboratories were among the earliest investigators to study the relationship between properties and structures, and to analyze the deposition process. However, most of this early work was done on material deposited at relatively low temperatures (below 1500°C) and the deposits formed were extremely thin. Very little additional work was done on the process until the early 1950's.

A large program was undertaken by the British Royal Aircraft Establishment in the late 1940's and early 1950's to study the pyrolytic graphite deposition process. Much of this work was presented in papers by Brown *et al.* (6–8). In the United States, the United States Navy Special Projects Office instituted a major program in 1958 to determine if massive shapes of pyrolytic graphite could be produced for missile heat shields and nose cones. Included in this program was a thorough study of the structure and properties of the material. The initial program was under the direction of the Lockheed Missile and Space Company, with major studies and production of products by the Raytheon Company, High Temperature Materials Company (now a division of Union Carbide), and General Electric Company. Shortly thereafter, the Air Force Materials Laboratory at Wright-Patterson Air Force Base, Ohio, undertook an extensive research program on pyrolytic graphite in which a number of companies and scientific laboratories participated. The Allegheny Ballistic Laboratories of the Hercules Powder Company also undertook a major program aimed at utilizing pyrolytic graphite as a rocket nozzle throat insert for the second stage of the Polaris A-3 missile. Both the Jet Propulsion Laboratory and the Marquardt Corporation carried out programs to develop free-standing rocket nozzle shapes of pyrolytic graphite.

The work done on the NERVA nuclear rocket by the Los Alamos Scientific Laboratory and later by the Westinghouse Astronuclear Laboratory in the late 1950's and early 1960's furthered development of pyrolytic graphite science.

Studies of material properties and deposition process variables are still being carried out by independent laboratories and by the major producers of the material. There are now at least four major producers of pyrolytic graphite within the United States and half a dozen or more independent laboratories where extensive research and materials testing work is being done. Much of the work now in process is devoted to producing new and improved structures and "alloys." Unfortunately, commercial applications for pyrolytic graphite are limited by its price of $40 or more per pound. It is now estimated that the annual production capacity in the United States is around 20 tons per year, with annual sales of around $2 million to $4 million. As far as the authors know there are no major production facilities outside of the United States. Possibly some exist in France and the iron-curtain countries; a review of available Soviet literature, however, indicates that the Soviets have not disclosed any extensive use of pyrolytic graphite.

II. Manufacturing of Pyrolytic Graphite

A. Basic Manufacturing Considerations

It is possible to manufacture pyrolytic graphite in commercial quantities in a wide variety of equipment. Certain essential elements are common to all equipment, however.

1. A Gaseous Source of Carbon

The most economical material for the deposition of pyrolytic graphite is natural gas, which normally contains 90 to 95% methane, 2 to 5% acetylene, and from 2 to 5% other hydrocarbons, nitrogen, and hydrogen, with very minor amounts of water vapor. Usually a very small fraction of a percent of mercaptan is added to the natural gas to give it a distinctive odor. Other deposition source gases (6) include bottled methane, propane, acetylene, higher hydrocarbons, and fluorcarbons. All of these are less economical to use than natural gas and do not produce any significant improvement in the deposition rate or the material produced.

Experiments have been carried out in which some of the lower boiling point fractions, such as the mercaptans and acetylenes, have been removed from the natural gas by passing the gas through an acetone dry-ice trap before it is introduced into the furnace. Usually a small amount of liquid is collected in these traps, but it has not been shown that going through the additional labor and expense of removing these hydrocarbons, and other less volatile materials, produces a pyrolytic graphite superior to that produced from untreated natural gas.

2. The Deposition Substrate

Once the hydrocarbon gas is selected, an appropriate substrate must be chosen. The deposition rate of pyrolytic graphite has been shown (5, 6, 9, 10) to vary sensitively with substrate temperature. Below 1500°C at low pressures the deposition rate is extremely slow (being measured in fractions of a mil per hour). Above 2250°C the deposition rate does not appear to be significantly increased by a further increase in temperature. Above 2500°C, considerable alteration of the structure occurs during deposition, and the material produced resembles single-crystal graphite to a much greater extent than it resembles commercial-grade pyrolytic graphite. This point will be more fully discussed in Section III, on the structure of the material.

It has not been found economical to deposit commercial quantities at low pressures above 2250°C, since considerable vaporization of graphite occurs from the substrate and other components in the hot zones. Considerable deterioration in the strength and other properties of the graphite hardware in the furnace is also evident. Low deposition rates preclude operating the substrate much below 1750°C. Therefore, most commercial pyrolytic graphite is deposited at substrate temperatures between 1750° to 2250°C.

When a relatively narrow substrate temperature range has been selected, it is then necessary to specify the type of substrate on which the deposition process is to be carried out, and to provide an economical method of heating.

The process of depositing pyrolytic graphite is carried out in a highly carbonaceous atmosphere by the thermal cracking of a hydrocarbon. Therefore, there are severe limitations on the type of substrate upon which the deposition process can be carried out. Certainly, materials that do not have adequate strength, or which melt, vaporize, or are chemically reduced below the deposition temperature, cannot be considered as usable substrates to produce a normal commercial-grade pyrolytic graphite. In some cases, it is possible to deposit on carbide-forming materials if the rate of carbide formation is lower than the rate of pyrolytic graphite deposition.

From a practical standpoint, only three or four materials can be considered satisfactory substrates for deposition. These are the refractory metals, refractory hard metal compounds (carbides, nitrides, and borides), and normal commercial polycrystalline graphite. The refractory metals, such as molybdenum, tungsten, and tantalum, are carbide formers and are expensive, as are the refractory hard metal compounds. Commercial polycrystalline graphite is, however, an inexpensive material; it is produced in tonnage quantities and in a wide variety of sizes, shapes, and grades. Therefore, both the process and the economics of the situation dictate that commercial polycrystalline graphite be used as the deposition substrate. Of course, once a sheet of pyrolytic graphite is obtained it can also be used as a substrate, but in general the cost is prohibitive. Unfortunately, the majority of polycrystalline graphite is quite porous and it is difficult to obtain a very smooth surface. This has a marked effect on the structure of the pyrolytic graphite that is deposited on this substrate, as discussed in Section III on microstructure and macrostructure. Pyrolytic graphite deposited on a substrate replicates the structure of the substrate. Therefore, a depression or a small bump on the substrate affects the structure, which in turn governs the material's mechanical properties.

The selection of the grade of graphite and its preparation as a substrate, therefore, strongly affects the quality of the pyrolytic graphite produced. Optimally then, we need a graphite of very low impurity content, very fine particle size, and as high density as possible. If we selected a graphite substrate on this basis alone, however, we would find three difficulties: (a) the cost of the graphite would be prohibitive, (b) the material might have inadequate thermal shock resistance and (c) graphites such as these would not be available in the size and range required. It is often necessary, therefore, to select much less dense graphites with larger particle sizes than optimal. This course provides low-cost sections that are large enough and have adequate thermal shock resistance to use in the manufacture of many of the pyrolytic graphite shapes required by industry today.

3. Control of Pressure, Temperature, and Gas Flow

Both the gas and the substrate having been selected, a method of heating the substrate to the desired temperature must be provided. Enclosing the substrate in a vessel permits the exclusion of air, and allows the control of pressure and flow rate of gas through the hot zone of the furnace. The exact pressure required for the deposition process is a function of the type of hydrocarbon used, the temperature of the substrate, and the geometry within the hot zone of the furnace. If we look at the deposition process as a cracking of a hydrocarbon we can better visualize how the process takes place. If we do not control the pressure, temperature, and other variables within certain limits, we will produce a powder deposit of soot or lampblack. The problem is to prevent the gas molecules from reaching in the gas phase forming large macromolecules of dehydrogenated hydrocarbons. As the particles grow they become large enough to precipitate out of the gas stream in the form of soot or lampblack, and loosely adhere to the hot wall. If they are controlled to deposit at the wall while still small—or better, to decompose at the wall—a pyrolytic graphite coating results.

Therefore, consideration must be given to controlling the number of collisions that occur in the gas phase compared to the number of collisions the gas makes with the deposition substrate. The mechanism of energy exchange between cold gas molecules and a hot surface has been studied by Meyer (9), Diefendorf (10), Brown et al. (6–8), and Voice (11), who have investigated the deposition process as well as the deposit. In studying the process variables, the pressure of the gases within the furnace must be measured, not in the hot zone, but in some more remote region. The gas mixture at such a point probably consists of a mixture of hydrocarbons and hydrogen or inert carrier gases (if they are used). Normally, the pressure measured is within the range from ½ to 50 torr (millimeters of mercury) without a carrier gas.

Generally, the method of obtaining this low pressure is to use mechanical vacuum pumps. If the partial pressure of the hydrocarbon gas is kept in the range of ½ to 50 torr, satisfactory deposition conditions can be sustained. A second method is to operate the furnace above atmospheric pressure and use an inert carrier gas such as helium, neon, or argon.

Since the number of gas-gas and gas-surface collisions that occur is critical, the product geometry and gas path length have a profound effect on the process operational pressure. A furnace in which the residence time of the gas is long cannot be operated at as high a pressure as a furnace with short residence time. Two furnaces essentially alike but in which the size scale factor is 1 to 5 cannot be operated at the same pressure. A small furnace can be operated successfully at a pressure of 20 torr without producing soot, for example, but a larger furnace could not operate successfully above 5 torr of pressure without producing large quantities of soot. The exclusion from the furnace of oxidizing elements, such as air, is also necessary, regardless of how the low pressure is obtained. Oxidizing elements are believed to affect deposition adversely by decreasing the size of the carbon species and preventing them from coalescing.

Practically, in the commercial production of pyrolytic graphite in large quantities, vacuum pumps rather than an inert carrier gas are invariably used. The cost of the carrier gas, whether argon, nitrogen, or helium, is such that economy requires a system for recovering and purifying the inert gas. This is quite expensive. Aside from equipment cost for recirculation small losses of even a few percent of carrier gas are economically prohibitive. Conversely, the techniques and costs of producing and maintaining vacuums in the range of ½ to 50 torr are very reasonable. Equipment is readily available on a commercial basis for carrying out this operation over a wide range of furnace sizes.

The vacuum pumps used are either the mechanical rotary oil type, the blower types backed up by mechanical rotary pumps, or molecular pumps. Sometimes four- or five-stage steam ejector pumps are also used. The selection of the types of pumps is to a large extent dictated by availability, furnace size, ease of operation and installation, maintenance, and initial cost. Normally, mechanical rotary oil pumps are used because of their relatively low cost and ease of operation. The blower-type pumps are somewhat more difficult to operate because of their limited pressure range and the large amount of dirt in the form of soot or lampblack that comes through with the gas stream. This tends to limit the pump life. The pump life can be extended by using filters; however, these cut down on the pumping capacity. Where a large number of furnaces or furnaces of a large size are to be used, it appears that steam ejector pumps are very satisfactory. To get to the desired pressures, however, multiple-stage ejector pumps are required. The initial investment in such a system is often greater than that for mechanical rotary oil pumps since a fairly large steam plant must be associated with the pump. If steam is needed for other plant purposes, and the total steam requirement for the steam ejector pumps is not too large, the cost of installing the boiler, maintaining it for long periods of time, and operating on a round-the-clock basis can be spread over a number of processes within a large plant. The total operating costs can then be considerably lower than for mechanical pumps. Mechanical pumps of the oil rotary type generally must be provided with some purification system whereby the oil is periodically, or continuously, removed from the pump and cleaned to remove dissolved hydrocarbons and sludge in the form of soot which collects in the oil. A number of commercial oil purifiers are available which permit this on a regularly scheduled or continuous basis.

4. The Furnace Shell

To provide a chamber in which the controlled pressure is to be maintained is in itself not a difficult engineering problem. Usually a simple low-carbon steel vessel suffices to contain the hot zone and to exclude the atmosphere. The steel shell is provided with a number of openings so that electrical power can be fed into the furnace. Openings are made so that gas can be introduced and removed, and windows are provided so that temperatures within the furnace can be measured. Provision is also made for the introduction of a gas-pressure measuring device and cooling water. Furnaces are readily available on the market from a number of

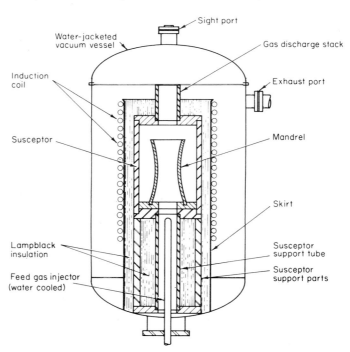

Fig. 1. Typical furnace construction for the production of pyrolytic graphite.

commercial establishments or can be designed and built readily. The general layout of a typical pyrolytic graphite furnace, showing schematically the construction of the vessel and the functions of the various openings, is given in Fig. 1. Since the hot-zone temperature is in the neighborhood of 2000°C, and considerable hot-gas radiation occurs from this area to the steel furnace, the vessel is usually of double-wall construction with water circulating between the two shells.

5. Heating Methods and Furnace Operation

The next selection must be the method of heating to obtain a substrate temperature of approximately 1750° to 2250°C. There are several opinions as to the best method of providing the required heating power. Usually, because of geometrical considerations and the wide range in complexity of shapes produced, it is not practical to heat the mandrels directly by resistance or induction heating. Rather, a large chamber is heated and the mandrels are subsequently heated, principally by radiation from the large chamber. The chamber itself is invariably constructed of graphite, and the mandrels placed within the chamber are also of graphite, as stated earlier. Since anything within the hot zone will receive a coating of pyrolytic graphite, it is generally necessary to prevent the hydrocarbon gas from coming in contact with the radiating walls of the chamber. These walls can thus maintain their shape and do not fracture from thermal stresses or deteriorate with time.

The two most commonly used methods of obtaining the desired temperature for the chamber are resistance and induction heating. The choice is determined by economics. The initial cost of induction equipment is usually higher than that for resistance heating. However, the power requirements for operating the induction furnace, once installed, are lower. Economic analysis indicates that for a furnace diameter up to 12 inches, induction and resistance heating are equally attractive. However, for diameters much above 12 inches, induction heating appears to many to be more economical. For induction heating a wide range of selections of frequency is possible. Furnaces have been operated at a frequency as low as 180 cycles and as high as the kilocycle range.

A major consideration in the use of induction heating is that the induction coil must be placed within the vacuum chamber and the coils insulated from each other and from the work. Care must be taken to insure that electrical discharge and puncturing of the induction coils does not occur. The induction coils are water cooled to withstand the high I^2R heat losses within the coils and to dissipate heat radiated from the hot zone.

The manufacturers of equipment utilized for vacuum melting and other vacuum processes usually specify the proper types of insulation to be used. Adjacent to the induction coil (or if the induction coil is not to be used, the graphite resistor) some material must be placed to serve as a thermal insulator. For induction heating, the insulation is placed between the induction coil and the chamber being heated (here called the susceptor) (Fig. 2). In all commercial installations, the insulation used today is lampblack or carbon felt. With resistance heating, it is necessary to prevent the hydrocarbon gas from coming in contact with the resistance element and depositing pyrolytic graphite on it. This is done by proper channeling of the gas flows (Fig. 2). The same holds true for the susceptor used in the induction heating process. As far as is known, graphite is used for all resistance elements and susceptors for the commercial production of pyrolytic graphite. A typical installation available on the market for an induction heating furnace is shown in Fig. 3. Often because of the heating-cooling, loading-unloading cycles of the furnace, it is practical to operate several furnaces from a single power supply. A large furnace with a hot zone of around 15 to 20 inches in diameter and from 25 to 35 inches long will have a furnace cycle somewhat as follows:

1. Loading time of the furnace is 2 to 12 hours, depending upon the complexities of the load.

2. Heat-up time is from 1 to 6 hours.

3. Deposition time depends upon deposition rate (usually about 5 to 10 mils per hour) and desired thickness.

4. Cooling time from the deposition temperature is from 10 to 30 hours.

It is generally more economical to keep the thermal insulation in the

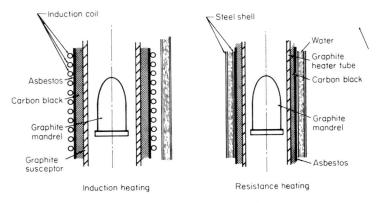

FIG. 2. Detail of furnace insulation.

Fig. 3. An induction furnace installation.

furnace at a minimum and provide excess power to operate the furnace rather than to have good insulation and operate at minimum power. If a balance is kept between the size of the power supply and the heating and cooling rate of the furnace, the most economical sizing of the power supply and the insulation will be realized. As far as it is known, most furnaces operate with a lampblack insulation thickness between 2 and 6 inches. For small furnaces, carbon felt has been found to be very satisfactory and much cleaner than lampblack. The cost of this material, however, has prohibited its utilization in very large furnaces.

Much of the graphite in a new furnace contains from ½ to 5% impurities, generally consisting of iron, silicon, magnesium, sulfur, and calcium. In a new furnace where the bulk of graphite weighs several tons, the impurities can represent a substantial number of pounds of material, which on evolution can adversely affect the deposition product. When the furnace is heated to 2000°C most of these impurities vaporize and condense in cooler sections of the furnace, such as the exhaust stacks, as indicated in Fig. 1. This material accumulates and forms low melting point carbides, which drip back into the furnace and splash against the deposition surfaces, creating very rough spots. When the hydrocarbon is introduced, these rough spots cause defects in the pyrolytic graphite. Before a new furnace is utilized for the deposition process, therefore, it is generally taken to a higher temperature than it will ever see during the actual deposition process. A system is provided to prevent the carbides and other materials from splashing back into the furnace. This generally is referred to as the furnace "bake-out" cycle and can take from 1 to 5 days to accomplish, depending upon the size of the furnace.

6. Instrumentation

Certain basic instruments are required to control the process variables during the deposition cycles. It has been found by experience that fluctuations in such variables as temperatures, pressures, and flow rates can cause the pyrolytic graphite produced to be markedly inferior. For example, if the pressure during a deposition run is changed, a microstructural change within the deposit will occur. With that microstructural change, a change in expansion coefficients and thermal conductivity from one portion of the deposit to the other will result. Usually the difference in these properties causes stresses due to differential contraction, and the material will cleave or fracture. To prevent this, and get a homogeneous deposit, it is necessary to control, within close limits, the temperature, pressure, and flow rate. It has been found desirable to control the pressure within 0.5 torr and to control the gas flow rates within 5% during the entire deposition period. Temperature usually must be held within ±20°C if major variations in the structure are to be avoided.

Many instruments for accomplishing this degree of control are available on the market. It is possible, for example, to continuously monitor and record furnace temperature with a radiation pyrometer, which will indicate the temperature in the hot zone during the period of operation. The exact temperature within the furnace can then be controlled either by a servo mechanism connected to the power supply or manually. It has been found that with large furnaces, the mass of graphite within the furnace is such that a large thermal lag occurs, and manual control is very satisfactory. As the size of the furnace decreases, the thermal lag decreases and it is more difficult (but not impossible) to maintain the temperature within the close limits by manual methods.

In addition to the need to control temperature, pressure, and flow rate, certain safety features of the furnace must be continuously monitored if serious and costly accidents are to be prevented. It is necessary, of course, to insure that a good supply of suitable process water is provided to the furnace, to cool the induction coils or the electrical contacts. It is also necessary to provide water to cool the vessel. In cases where the gas is introduced through water-cooled injectors that are exposed to radiation from the high temperature within the hot zone, adequate cooling for these is also essential.

It has been shown analytically that at normal furnace pressures, the quantity of hydrocarbon present in the furnace, if reacted with oxygen at the temperature of the hot zone, would not generate sufficient pressures to constitute an explosion hazard. The major source of explosion hazard arises from the possibility of a large water leak within the furnace and a resultant steam and hydrogen explosion. A number of such mishaps have occurred in the past. However, to the author's knowledge, no one has ever been seriously injured as a result. Usually some equipment is lost—either the induction coil is melted, or the studs through which the electrical connections to the resistance elements are made are destroyed. Most furnaces built today are provided with water-flow alarms, water-temperature alarms, auxiliary backup water systems, etc., so that if pressure rises within the furnace, water is shut off automatically and a major steam explosion cannot occur.

Within the limits of availability of large graphite parts and high capital investment, the larger the pyrolytic graphite furnace the more economical it is to operate.

The largest furnaces now in commercial operation have hot zones approximately 72 inches in diameter and about 72 inches in length.

B. Deposition Process

Although the first patent for a process to make a nonporous highly oriented graphite (pyrolytic graphite) was issued around 1880, no real attempts were made to understand the fundamentals of the deposition process until the mid 1900's. The kinetic mechanism, however, was well known long before the 1950's, and hydrocarbon decomposition was used extensively for the manufacture of lampblack and other types of carbon. The primary application for pyrolytic graphite in 1950 was in thin film graphite resistors. In 1951, as a result of the work of Grisdale and his associates (5), a detailed analysis was undertaken of the formation of low-temperature thin films of pyrolytic graphite. Although the deposition mechanisms were similar to those of 1750°C deposits, these low-temperature (1300°C) films were quite different from what we now call pyrolytic graphite. These investigators found that both the nature of the deposit and the gas decomposition are temperature dependent at a given pressure. They were able to show experimentally that the pyrolytic films consisted of minute crystal packets composed of parallel plane sheets of carbon atoms in hexagonal arrays as in single-crystal graphite.

The pyrolytic films differed from conventional single-crystal graphite in an important respect. Whereas in the layers of graphite the atoms lie one above the other in a definite geometric pattern, in pyrolytic films the major crystallographic order is only within the layer planes themselves, and the layer planes, except for being roughly parallel, are oriented randomly with respect to each other. This is termed a "turbostratic structure."

The mechanism for the deposition of these films was believed to proceed along these lines: The gas is introduced into the hot zone of the furnace and the methane molecules react to produce hydrogen, diacetylene and acetylene (5, 9). This is followed either by the growth of highly complex hydrocarbons of higher and higher carbon content or by the decomposition of hydrocarbons with the continuous stripping of the hydrogen from the macromolecules. These macromolecules are deposited on the surface, and once at the surface, continue to give off hydrogen. It has been found that as the temperature of deposition increases, the amount of residual hydrogen in the deposits decreases. It is now known that for depositions carried out at much higher temperatures than those used by Grisdale et al. (5), the amount of hydrogen left in the deposits is extremely low, normally less than 1 part per million. From a purely thermodynamic standpoint, the pyrolysis of methane to produce carbon and hydrogen is well understood. Considering the reaction $CH_4 = C + 2 H_2$, Table I (12) shows the equilibrium constant for the pyrolysis of methane. Here $K_p = P_{H_2}^2 / P_{CH_4}$. As can be seen from this table, at a temperature of approximately 800°K (1000°F), the free energy of formation of methane, ΔF, becomes positive, indicating that the reaction tends to proceed to the right, with free hydrogen and carbon as the more stable

TABLE I

THE EQUILIBRIUM CONSTANT FOR THE PYROLYSIS OF METHANE (12)

Temperature, °K	Log K_p	K_p
0	$-\infty$	0
250	-11.3534	
298.16	-8.8985	
300	-8.8184	
350	-6.9393	
400	-5.4899	0.000003
450	-4.3558	0.000037
500	-3.4273	0.000374
600	-2.0004	0.009991
700	-0.9529	0.1111
800	-0.1560	0.7080
900	$+0.4881$	3.077
1000	$+1.0075$	10.18
1100	$+1.4345$	27.19
1200	$+1.7936$	62.17
1300	$+2.1006$	126.1
1400	$+2.3638$	231.1
1500	$+2.5923$	391.1

TABLE II
EQUILIBRIUM COMPOSITION OF THE PYROLYSIS OF METHANE AT VARIOUS
TEMPERATURES AND PRESSURES (12)

Temperature, °K	K_p	Log K_p	x/a		
			$P = 0.1$ atm	$P = 1$ atm	$P = 5$ atm
650	0.05	−1.301	0.333	0.111	0.0499
690	0.1	−1.000	0.447	0.156	0.0705
	0.2	−0.609	0.577	0.216	0.0995
	0.3	−0.523		0.264	0.121
	0.4	−0.398	0.707	0.301	0.140
	0.5	−0.301		0.333	0.156
	0.6	−0.222	0.774	0.361	0.170
800	0.7	−0.155		0.386	0.184
	0.8	−0.0969	0.816	0.408	0.196
	0.9	−0.0458		0.428	0.207
	1.0	0.0000	0.845	0.447	0.218
	2	0.301	0.913	0.577	0.301
890	3	0.477		0.655	0.361
	4	0.602	0.953	0.707	0.408
	5	0.699	0.962	0.745	0.447
	7	0.845	0.972	0.797	0.509
1000	10	1.000	0.980	0.845	0.577
	13	1.114	0.985	0.875	
	16	1.204	0.987	0.894	0.667
	20	1.301	0.990	0.913	0.707
	25	1.398	0.992	0.928	0.745
1100	30	1.477		0.939	0.775
	40	1.602	0.995	0.953	0.816
	50	1.699	0.996	0.962	0.845
	100	2.000	0.998	0.980	0.913
1370	200	2.301	0.999	0.990	0.953
1550	400	2.602	1.000	0.995	0.976
	800	2.903		0.998	0.988
	1,600	3.204		0.999	0.994
	4,000	3.602		1.000	0.995
	10,000	4.000			0.999
	20,000	4.301			1.000
	50,000	4.699			

NOTE: This table is based on methane alone and does not take into consideration the presence of other hydrocarbons. Consequently, because of uncertainty in the hydrocarbon species involved above 1000°K, the quantitative thermodynamic treatment in that range is imprecise. For example, in place of a maximum of 0.005 atm of methane at 1550°K and 1 atm pressure, there may be about 0.08 atm of C_2H_2; 0.005 atm of C_3H_5; 0.1 atm of CH_4, and 0.07 atm of C_6H_6 (13).

phases. (Natural gas contains some hydrocarbons other than CH_4 which influence the equilibrium concentration of CH_4, particularly at temperatures above about 1000°K.)

In Table II (12), showing the equilibrium composition of the pyrolysis of methane, we see the effect of temperature on the breakdown of methane to produce carbon and hydrogen. Here the quantity x/a represents a ratio of the number of moles of carbon at equilibrium to the number of moles of CH_4 added initially. From this table, it can be seen that at temperatures around 700°K (427°C) and a pressure of 5 atm the reaction is at equilibrium with very little carbon. At a pressure of 0.1 atm the reaction is very highly favorable toward the formation of free carbon. At temperatures of interest for pyrolytic graphite deposition at less than 0.1 atm pressure, the reaction is essentially 100% complete. The mole fraction of methane in equilibrium with carbon is less than 1% under these calculated conditions. Samples of gases coming from pyrolytic furnaces have shown that 50 to 85% of the carbon is removed from the methane that is fed into the furnace. This would indicate that residence time in the furnace, kinetic conditions, and presence of other hydrocarbon species control the deposition process rather than the simpler calculated equilibrium conditions.

As the temperature of the deposition increases the size of pyrolytic layer particles increase. However, it is not known whether this increase results from the growth of the particles after deposition or whether the higher-temperature particles are bigger than those deposited at lower temperatures. For example, Grisdale et al. (5) found that the particles were usually less than 50 Å in diameter. Brown and Watt (6–8) found that the particles were from 50 to 200 Å in size. Recent work by others (11,14,15) using line-broadening X-ray techniques indicate a wide range in particle sizes within any deposit. The largest of these particles have diameters up to 500 Å. If the deposition is carried out above 2500°C, the particles grow out of the angstrom size range and are expressed in macrodimensions. At temperatures of 2750°C and above, the physical boundaries of the part determine the size of crystallites.

To review, if the temperature is in excess of 600° to 700°C, equilibrium thermodynamic considerations dictate that some free carbon will be formed. As the temperature is increased, the amount of methane in equilibrium with carbon decreases. Depending upon the pressure, the carbon formed can be produced in several ways. It can be produced as large molecular species containing small amounts of residual hydrogen. As we vary the gas temperature, pressure, or residence time in the furnace, we find that the gas phase reaction can become more important than the heterogeneous reaction at the wall, resulting in the formation of soot particles. So that a dense coherent deposit may be formed on the substrate, this sooting condition is prevented from occurring by reducing the pressure. Ideally the gas-phase reaction is minimized, limiting the size of the carbon particles that reach the deposition surfaces. The temperature at which methane shows appreciable decomposition is quite low, the decomposition being complete at temperatures in the neighborhood of 1200°C and at pressures of 0.1 atm or less. Therefore, the actual rate at which pyrolytic graphite is deposited is not primarily determined by the temperature.

Two factors determine the rate at which pyrolytic graphite is deposited within a furnace: the quantity of carbon contained in the gas and the residence time of the gas within the furnace. The latter is governed by reaction tube length and the flow rate. Here we have two competing effects; for example, we could flow a small amount of gas into the furnace at less than 0.1 atm so that most of the carbon would be stripped from the methane, or we could force a large quantity of gas flow through the same furnace. In the latter procedure, the residence time within the furnace would be very short and little decomposition of methane would occur. In the former procedure, we would have a low deposition rate because less carbon would be available for deposition. In other words, both low mass flow and short residence times of the gases in the furnace lead to low deposition rates.

The optimum conditions occur with a balance between the quantity of hydrocarbon gas entering the furnace and the time allowed for the gas to decompose. The geometry of each furnace governs the optimum decomposition conditions. An additional variable is the changing composition of the gas in the furnace. Initially, the gas is quite cold and very rich in hydrocarbon. As the temperature of the gas rises, gas-gas and gas-wall collisions occur, carbon is removed, and the composition of the remaining gas changes. The deposition from the source gas usually reaches a maximum in a local area of the furnace. Here the gas is quite rich in hydrocarbon, gas-wall collisions are occurring at a high rate and we have the maximum deposition rate. Further along the furnace the number of gas-gas collisions predominates. Instead of pyrolytic graphite, small quantities of very fine soot form in the furnace and become incorporated in the deposit.

Figure 4 is a deposition profile taken within a furnace several meters high and approximately a meter in diameter. The exact profile of the deposit can be changed by changing the pressure within the vessel, or by adding hydrogen, which changes the equilibrium conditions within the furnace. Hydrogen (one of the products of the reaction) tends to drive the reaction toward the stabilization of the hydrocarbon, thereby

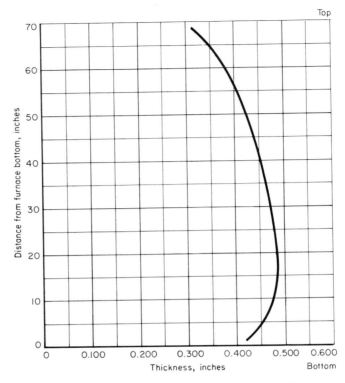

FIG. 4. Typical thickness profile for pyrolytic graphite produced in a 72-inch length hot zone.

slowing carbon deposition. The slowing of the carbon deposition process, and the depletion of the carbon in the gas, control the deposition profile within the furnace. Thus, the use of a diluent gas maintains the carbon concentration at a more constant value along the reaction wall.

It is sometimes desirable during deposition to use conditions that allow a small amount of very fine soot to be produced and incorporated in the deposit. This soot inclusion leads to a continuously nucleated microstructure that significantly affects the mechanical and physical properties of the pyrolytic graphite. This point will be more completely covered in later sections.

The question of the method of controlling the degree of crystal structure perfection or preferred orientation in the deposition has often been raised. As a matter of fact, the manufacturers exercise very little control over the degree of perfection or preferred crystal orientation. Once the temperature and pressure of deposition have been selected, and the substrate condition dealt with, it is virtually impossible to get any other degree of perfection or orientation than is characterized by these two conditions. No one has satisfactorily explained in detail the manner in which pyrolytic graphite is deposited (10,11). Many investigators have attempted to explain how the deposit forms with the majority of crystallites oriented with the hexagonal layer planes parallel to the deposition surface. It would be very helpful under certain circumstances to be able to make the pyrolytic graphite so that the basal planes are perpendicular to the deposition surface rather than parallel to the surface. As far as it is known, this has not been accomplished. Investigators have attempted to do this by various means, such as applying magnetic or electrostatic fields, and have found these methods completely ineffective. The theory that the crystallites are deposited somewhat like snowflakes, which have a geometrical anisotropy, and does not allow the molecules to stand on edge has certain fundamental weaknesses. From X-ray diffraction evidence it would appear that the ratio of the diameter of the crystallites to their thickness is only a factor of 2 to 5, which is low compared to the much greater snowflake anisotropy. No complete explanation can be made of why at least some significant fraction of the deposit would not orient itself with the basal planes perpendicular to the deposition surface.

In making flat plate in a pyrolytic graphite furnace, it has been found almost impossible to produce a straight piece of material having uniform thickness over its entire length. The usual curvature of plate is generally referred to in the industry as "bow" and arises from several factors:

1. When plate is produced, we essentially have a closed shape within the furnace, since all surfaces of the starting substrate are coated and the edges of the plate naturally touch each other in the hot zone. As the deposit cools from the deposition temperature, the differential between thermal contraction of the substrate material in the thickness and surface directions imposes stresses on the deposit that tend to deform the plate and cause it to bow.

2. Since the deposited material is not uniform in thickness, thermal contraction is not uniform, and this introduces further stresses in the deposit.

3. Finally, upper layers of the deposit are formed on expanded earlier sublayers, thereby keeping the sublayers in the expanded state. As the total deposit cools, the upper layers are compressed by the shrinking lower layers and the plate bows.

Large plate which is now produced in dimensions of 16 by 60 inches normally has a bow in the width direction ranging from ¼ to 1½ inches, and in the length direction from ½ to 4 inches. The amount of bow is very dependent on the thickness of the plate produced and on the microstructure.

It is estimated that within the United States today there are approximately 10 furnaces in operation with hot zones in the range of 20 to 72 inches in diameter and 30 to 70 inches in length. These are mainly used for pyrolytic graphite plate production, and it has been estimated that at full capacity they would have production capability of about 20 tons of plate per year. In addition to this, there are from 30 to 50 furnaces having hot zone diameters of around 8 to 20 inches. These are used mainly for the production of various composites of pyrolytic graphite, free-standing shapes and numerous small parts. Samples of large plate and free-standing parts are shown in Fig. 5.

III. Structure

Understanding of the atomic structure of pyrolytic graphite is necessary to understanding of the properties of the material. Therefore, this

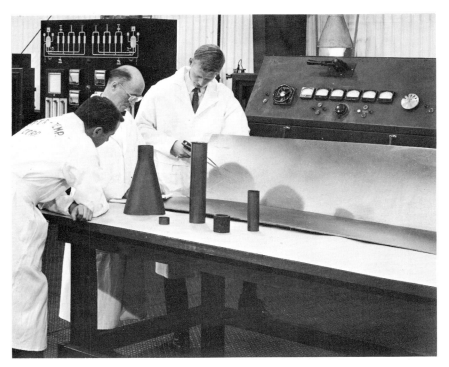

FIG. 5. Typical pyrolytic graphite plate and free-standing parts.

section on structure will deal first with the material's gross atomic arrangement, then its atomistic crystal structure, and finally its macroscopic polycrystalline organization. The structure of graphite in general has been covered by Shobert (4).

The atomic arrangement of a unit cell of graphite (Fig. 6) is hexagonal, with an orderly stacking of the hexagonal planes. The crystallographic directions A, B, and C shown in the figure will be referred to throughout this discussion. From bond energy considerations we know that the bonding between the atoms of the carbon within the plane (AB direction) is of the covalent type and therefore very strong. However, all of the bonding energy is utilized in forming these bonds with the atoms in the plane and no valency bonding is left to give high bond strength between the hexagonal planes of atoms. The only bonding that exists between the planes is of the residual type, sometimes referred to as Van der Waals bonding. The different types of bonding within the graphite lattice are important to the very high anisotropy of graphite and pyrolytic graphite.

The high bond energy within the planes means that the material is strong and has good electrical and thermal conductivity in all directions within the plane. Between the AB planes (C direction) the shear modulus for a single crystal of graphite is estimated to be only about 700×10^3 psi and Young's modulus about 5.5×10^6 psi. Along the plane (AB direction) Young's modulus is close to 140×10^6 psi, which is higher than that

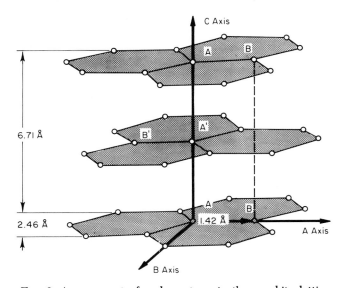

FIG. 6. Arrangement of carbon atoms in the graphite lattice.

of any other known material. The loose bonding in the C direction unfortunately does not allow us to take full advantage of the material's high elastic modulus. Therefore the material has very low mechanical strength and low thermal and electrical conductivity in the C direction.

So far we have been discussing bulk single-crystal graphite and not specifically pyrolytic graphite. Pyrolytic graphite differs in certain fundamental respects. The temperature of deposition, for example, greatly affects the degree to which the structure approaches that of single-crystal graphite. Pyrolytic graphite deposited at temperatures of about 1000°C shows little resemblance in crystal structure to Fig. 6, but more closely resembles the structure of a glass, with very little ordering of the atoms. As the deposition temperature increases, the degree of ordering increases. At 1800°C the degree of ordering of the AB plane atoms approaches 100%; however, the stacking of the hexagonal layers (C direction) is not as shown in Fig. 6. The AB layers, instead, appear to be stacked in a random fashion in the C direction.

As deposition temperatures of 2400° to 2500°C are approached some C-direction ordering begins to appear between the planes, and the structure resembles that of the graphite in Fig. 6. The spacing between

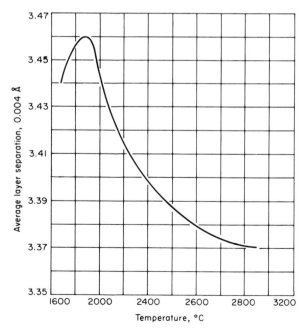

FIG. 7. Effect of deposition temperature on pyrolytic graphite layer plane separation (~004).

the layers of single-crystal graphite or C_0 is 3.35 Å. Pyrolytic graphite deposited at about 2000° to 2200°C has a C_0 spacing of about 3.44 Å. That is, the layers are spread farther apart than they would be in bulk or single-crystal graphite because of disorder in stacking of the layers one above the other during deposition. A typical curve of the lattice spacing of pyrolytic graphite as a function of the deposition temperature is given in Fig. 7.

Figure 8 shows the change in lattice spacing of pyrolytic graphite heated to an elevated temperature. The increased thermal energy apparently causes rearrangement of the atoms within the lattice. The structure of the pyrolytic material slowly approaches that of bulk graphite or a single crystal until at 3000°C, it has many single crystal characteristics. The density increase that accompanies this lattice parameter change is substantial. Thus, as the average lattice spacing drops from 3.44 to 3.35 Å, the density increases from 2.20 to 2.26 gm/cc. Typically, pyrolytic graphite manufactured in large quantities is made at temperatures in the range of 1800° to 2200°C. At these temperatures the material has a density of approximately 2.20 ± .02 and the lattice spacing C_0 is 3.44 Å.

The variation in C-direction stacking of planes between bulk and pyrolytic graphite does not fully account for their property differences. How the layer planes are arranged on a microscopic scale within the

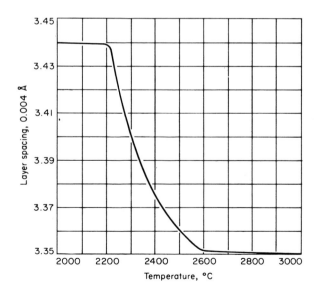

Fig. 8. Typical changes in layer spacing (~004) for pyrolytic graphite made at 2000°C and annealed for 4 hours at elevated temperatures.

pyrolytic graphite structure is also important. From electron micrographic and X-ray diffraction work (*15,16*), the structure of pyrolytic graphite can be described in terms of small packets of crystals with the intracrystalline atoms arranged in sheets. The individual crystallites assemble into polygonal zones or grains connected by tilt-boundaries and having the appearance of wrinkled sheets. This wrinkling has a marked effect on structure and properties and was observed in the electron micrographic work of Stover (*15–18*), Kotlensky and Martens (*19*), Brown and Watts (*6*), and others.

In Fig. 9, a reproduction of an electron micrograph, the nature of the wrinkled sheets of pyrolytic graphite is shown. Figure 10 shows the schematic diagram of the edge view of the same type of structure, as visualized by Kotlensky and Martens (*19*). From line-broadening work by the above authors and by Pappis and Blum (*20*) and others (*21,22*), the basic crystallite size is shown to be a function of the deposition temperature or the subsequent annealing temperature.

It has been seen that pyrolytic graphite is not a single invariant material, but has a structure and related properties that are completely dependent upon the manufacturing process. When made at very high temperatures, pyrolytic graphite is apt to resemble more closely single-crystal graphite.

The particular deposit shown in Fig. 9 was made at about 2200°C. Now, if we were to take this deposit to a higher temperature and subject it to tensile elongation along the basal plane, or if we were to deposit material at a higher temperature, we would find fewer wrinkles in the planes. The deposit in Fig. 11 was made at 2200°C and subsequently

Fig. 9. Electron micrograph of as-deposited pyrolytic graphite. Courtesy Jet Propulsion Laboratory.

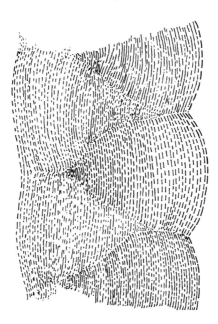

Fig. 10. Schematic diagram of the edge view of a plate of pyrolytic graphite showing arrangement of crystallite.

heated to 3000°C long enough to accomplish transformation to a single-crystal structure. As can be seen by comparing Fig. 9 and Fig. 11, the wrinkles in the sheet have been eliminated and the crystallite size has changed by probably two orders of magnitude.

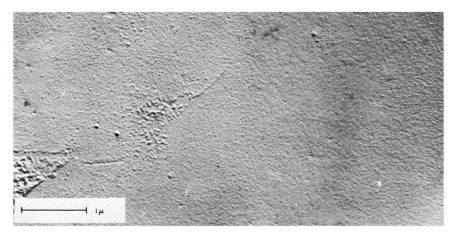

Fig. 11. Electron micrograph of pyrolytic graphite after heating to 3000°C.

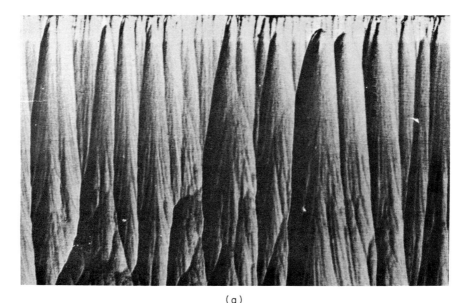

(a)

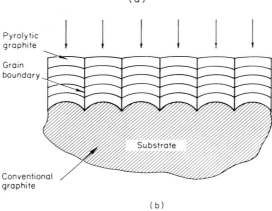

Pyrolytic graphite

Grain boundary

Substrate

Conventional graphite

(b)

FIG. 12. Wrinkling in pyrolytic graphite. (a) Pyrolytic graphite deposited on commercial graphite (65); (b) the effect of substrate surface roughness on the structure of pyrolytic graphite (23).

A good correlation between the theoretical mechanical properties and the actual properties has been found by Stover (16). Two factors are believed to play an important part in characterizing the deposit as wrinkled sheets of atoms. First, the basic crystallite that forms during the deposition process has a certain degree of shape anisotropy. It is known that the crystallites are between 50 and 200 Å in diameter and about 30 and 100 Å in thickness, the exact size depending upon the deposition parameters. Because of the dimensional differences between thickness and diameter of the crystallites, deposits are somewhat disarrayed with respect to one another, or wrinkled. Second, the substrate on which deposition occurs influences wrinkling. As far as the present authors know, no study has been made of pyrolytic graphite deposited on microflat substrates. A deposit made on a conventional graphite substrate is shown in Fig. 12. Apparently the surface is anything but smooth. Much of the wrinkling of the sheets and crystallites occurs because the substrate is not atomically flat. This relationship between the substrate and the microstructure was examined in great detail by Coffin (23). The effect of the substrate in producing curved or wrinkled planes is also illustrated in this figure. Although this schematic representation was primarily designed to show how large cones are developed in pyrolytic graphite by surface irregularities, it can be applied on a much finer scale to account for the wrinkling of the structures shown in Fig. 11. The substrate on which the deposition occurs, and which is replicated almost on an atom-by-atom basis, therefore, accounts for the surface of the deposit. In Fig. 12 another important property of pyrolytic graphite is shown. It can be seen that there is a correlation between the degree of bending of the basal planes and the amount of wrinkling that occurs. Basal-plane bending is a function of the distance of the plane from the defect in the substrate. This means that the material has a graded structure through its thickness. Since the material properties are anisotropic, the fact that the basal planes are tilted more when they are closer to the substrate means that planes next to the substrate are less anisotropic than those located some distance away. Further, the material first deposited remains hot during the entire deposition, whereas that last deposited has not been hot as long. The first-deposited material, therefore, has a tendency to anneal and lose wrinkles, a process which counteracts the tilt of basal planes at the substrate. Stover (16) has shown how the degree of preferred orientation and the properties vary through the thickness of the deposit. Changes that occur on depositing or subsequently heated pyrolytic graphite at elevated temperatures were studied by Richardson and Zehms (14) and in the earlier works of Guentert and Prewitt (21).

IV. Properties

A. MECHANICAL PROPERTIES

1. Compressive Strength

In contrast to what might be expected, compressive strength of pyrolytic graphite is higher in the C direction than in the AB direction. In

Fig. 13 the temperature dependence of compressive strength for both orientations is given. It should be noted that because the compressive strength in the AB direction is 50% lower than compressive stresses obtained in flexure testing (see Section IV,A,4), the validity of the data is under question. Specifically, below 2200°C flexure failures should have been compressive rather than tensile as observed, if these data were valid. Factors to be considered in accounting for the difference included (a) weak interlaminar bonding causing buckling under compressive loading, (b) end effects that may influence the relation of stress distribution to compressive stress, and (c) Poisson's ratio effect across the C-direction planes. Poisson's ratios for the A and C planes are 0.18 and 1.0, respectively. Under an axial compressive stress, e, the AB plane will contract by $0.18e$ and the C plane will expand by e. Calculations show that at the compressive limit, the strain across the planes (C direction) is of the same magnitude as that which would cause failure by axial tension; thus an interlaminar tensile failure is probable. As indicated by the composite of data from Raytheon and Lockheed in Fig. 13, the C-directional strength (shown to 800°C) increases with temperature and would probably continue to increase linearly to 2200°C. Fracture is typically brittle and shows the familiar centrally squeezed biconical characteristic.

Like the tensile strength, the compressive strength is sensitive to the cone size. Large cones and nodules cause stress concentrations that enhance any failure mechanism. Room-temperature compressive strength decreased in one experiment from 66,000 psi to 36,000 psi as the cone

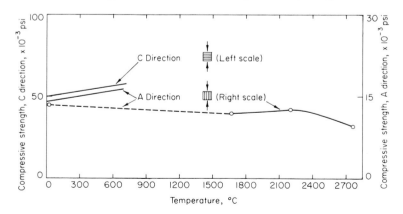

FIG. 13. Compressive strength of pyrolytic graphite as a function of temperature. (Composite of data from Raytheon Company, Pyrographite Research and Development, Final Rept., Subcontract PO18-2259, April 1960–August 1961; and Lockheed, LMSC Pyrolytic Graphite, Final Rept. No. LMSC-801376, **1** and **2**, June, 1962.)

size increased. Finally, and again in contrast to results for tensile strength, continuously nucleated microstructures exhibit 35% lower compressive strengths than substrate-nucleated microstructures.

2. Tensile Strength

The high strength-to-weight ratio for tensile strength in the A direction is shown in Fig. 14, as compared with other high temperature materials (24).

In the C direction, the tensile strength of pyrolytic graphite is lower than in the AB direction by a factor of 10 to 30 (20). Because of the difficulty of obtaining samples of suitable thickness, accuracy in the measurements has not been sufficient to identify this relationship more closely (25). Results obtained from notched specimens (Fig. 15) suggest a decrease with increasing temperature up to about 2000°C.

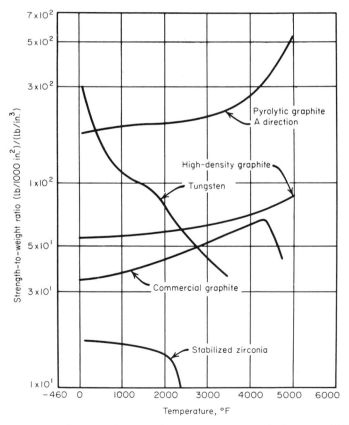

FIG. 14. Tensile strength-to-weight ratio comparison, A direction (24).

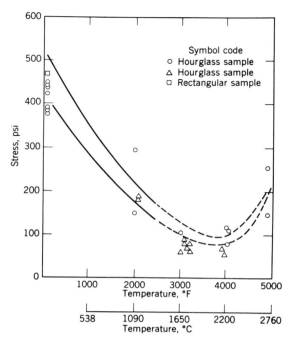

Fɪɢ. 15. Ultimate tensile strength vs. temperature, pyrolytic graphite, C direction. Notched specimens tested at different times (20).

The ultimate tensile strength of pyrolytic graphite in the AB direction is between 5 and 10 times greater than that of conventional pitch-coke graphites (26). As Fig. 16 shows, the increase in tensile strength with temperature accompanies an increase in the deformation prior to fracture (27). It would probably be overly optimistic to believe that the tensile strength of the pyrolytic graphite will increase much beyond 2500°C without failure (25).

As anticipated by the statements in the section on compressive strength, continuously nucleated microstructures exhibit higher tensile strengths than those of the substrate-nucleated microstructures.

3. Moduli of Elasticity and Rigidity and Poisson's Ratio

Table III presents the elastic constants of pyrolytic graphite as determined by numerous techniques.

The C direction modulus of elasticity and the Poisson's ratio were measured by standard strain gage techniques. The modulus of rigidity, G, in the AB direction was determined by the elastic equation relating modulus of elasticity, E, and Poisson's ratio μ (27).

$$G = \frac{E}{2(1 + \mu)}$$

The modulus of rigidity in the C direction was calculated by assuming that the difference between E in pure tension and E in beam deflection is due to a shear component in the beam deflection method. Since the shear deflection in the parallel orientation was approximately 15% of the total deflection, a shear modulus of 0.1×10^6 psi was calculated.

The negative value of the Poisson's ratio in the AB direction, although seldom encountered in engineering materials, is a necessary consequence of the high contraction in the C direction. Lockheed Missile and Space Company compression results (obtained in tension) are slightly higher.

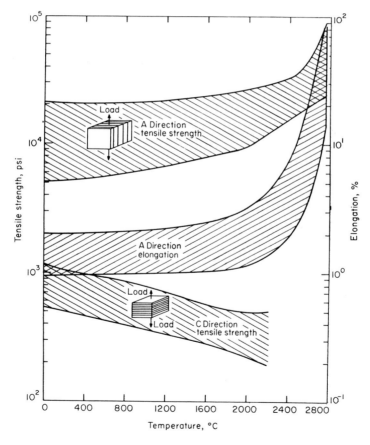

Fig. 16. Tensile strength and elongation of pyrolytic graphite as a function of temperature (26).

TABLE III

ELASTIC CONSTANTS FOR PYROLYTIC GRAPHITE (2100°C DEPOSITION)

Elastic constant	Symbol and direction	Mode	Value of constant	Remarks
E_a	Modulus of elasticity	Pure tension	4.29×10^6 psi	$E = \sigma/\epsilon$
		Tension, bending (average of $\parallel$ and $\perp$)	4.38×10^6 psi	$E = (Mc/I)/\epsilon$
		Compression, bending (average of $\parallel$ and $\perp$)	4.19×10^6 psi	$E = (Mc/I)/\epsilon$
		Beam deflection, $\parallel$ orientation	3.35×10^6 psi	$E = Pl^3/48Iy$
		Beam deflection, $\perp$ orientation	3.80×10^6 psi	$E = Pl^3/48Iy$
		Dynamic	3.20×10^6 psi	$E \propto \omega^2$
E_c		Pure tension	1.55×10^6 psi	$E = \sigma/\epsilon$
		Pure compression	1.58×10^6 psi	$E = \sigma/\epsilon$
G_a	Modulus of rigidity	Translaminar	2.55×10^6 psi	Calculated from $G_a = E_a/[2(1 + \mu_{ab})]$

		0.10×10^6 psi	Calculated from
G_c	Interlaminar		$G_c = K \dfrac{E_{a\text{-tension}} \times E_{a\text{-bending}}}{E_{a\text{-tension}} - E_{a\text{-bending}}}$ (K is a constant.)

Poisson's ratio (tension)

μ_{ab}	Transverse strain in planes due to stress along planes	-0.15		$\mu_{ab} = -\epsilon_y/\epsilon_x$
μ_{ac}	Transverse strain across planes due to stress along planes	0.90		$\mu_{ac} = -\epsilon_y/\epsilon_x$
μ_{ca}	Transverse strain along planes due to stress perpendicular to planes	0.35		$\mu_{ca} = -\epsilon_y/\epsilon_z$

Stress-strain relationships in pyrolytic graphite:

$$\epsilon_a = (1/E_a)(\sigma_a - \mu_{ab}\sigma_b) - (\mu_{ca}/E_c)\sigma_c$$
$$\epsilon_b = (1/E_a)(\sigma_b - \mu_{ab}\sigma_a) - (\mu_{ca}/E_c)\sigma_c$$
$$\epsilon_c = (-\mu_{ca}/E_c)(\sigma_a + \sigma_b) + (1/E_c)\sigma_c$$
$$\gamma_{bc} = (1/G_c)\tau_{bc}$$
$$\gamma_{ac} = (1/G_c)\tau_{ac}$$
$$\gamma_{ab} = (1/G_a)\tau_{ab} = [2(1 + \mu_{ab})/E]\tau_{ab}$$

where E, G, and μ are elastic constants, ϵ is normal strain, γ is shear strain, σ is normal stress, and τ is shear stress. Subscripts a, b, and c refer to the principal directions of pyrolytic graphite.

The designer's attention should be called to the Poisson's ratio in the AC direction, which does not appear in the stress-strain equations, but which can cause failure (see Section IV,A,1 on compressive strength) (24).

The modulus of elasticity is probably most valid when measured by the pure tension method. Table III includes three constants determined by beam methods; of these the strain method has the advantage of yielding both tensile and compressive data simultaneously. The strain method values are in error, however, owing to the shift of the specimen's neutral axis. The deflection moduli show the relative effects of the shear modulus for the two orientations (27).

No data have been reported on the effect of temperature on modulus of rigidity or Poisson's ratio. Figure 17, however, shows the temperature dependence of E as determined in tension, compression, and deflection tests.

The dynamic modulus for pyrolytic graphite in the AB direction is approximately 10% lower when continuously nucleated than when nucleated on the substrate. A slight increase in E between 120° and 370°C has been noted for both microstructures. In the absence of relaxation, linear decrease is expected for polycrystalline materials. Here, the decrease does not become linear until the temperature exceeds 950°C. In pyrolytic graphite, the lack of linearity between 370° and 950°C may be caused by microscopic viscous flow between conical crystallite boundaries (25). There appears to be little significant effect of microstructure on E.

As shown in Table IV, however, deposition temperature exerts con-

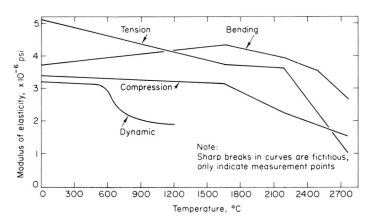

Fig. 17. Modulus of elasticity of pyrolytic graphite as a function of temperature as measured by various techniques (27).

TABLE IV
EFFECT OF DEPOSITION TEMPERATURE ON ELASTIC CONSTANTS

	Deposition temperature			
	1700°C	1900°C	2100°C	2300°C
Modulus of elasticity ($\times 10^{-6}$ psi):	3.34	5.60	4.29	4.10
Modulus of rigidity ($\times 10^{-6}$ psi):	1.8	3.0	2.5	2.4
Poisson's ratio, (μ_{ab}):	−0.09	−0.07	−0.15	−0.16

siderable control over E. Here again, the shear modulus was calculated from E and μ (27).

If valid moduli are to be obtained from Hooke's law equations, the principal stress directions must be along and across the planes of pyrolytic graphite. If they are in any other direction, new elastic constants must be used. Note that E is very sensitive to orientation. A 10° deviation from the planes results in a 50% decrease in stiffness. Because of the low interlaminar shear modulus, a minimum stiffness of less than E in the C direction is realized at 50°. This angle maximizes the elastic deformation due to interlaminar slip (27).

4. Flexure Strength

The ultimate flexural strength as a function of temperature is shown in Fig. 18. Flexural strengths for all three orientations are represented.

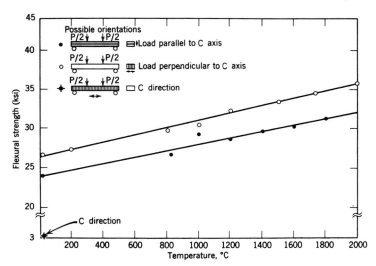

FIG. 18. Ultimate flexural strength of pyrolytic graphite as a function of temperature (20).

Apparently the flexural strengths parallel to the depositional planes are equivalent, but the strength parallel to the planes of the crystallite (held together largely by van der Waals' bonding) is very weak (20).

In the brittle temperature range, the failure mode in flexure of the parallel orientation is similar to that of wood [initial tensile failure of the outer fiber followed by an interlaminar tensile failure (28)]. The smooth fracture surface resulting from the perpendicular orientation indicates a simple tensile failure throughout the beam thickness. Above the brittle range, it is believed, the initial yield is in compression and the ultimate failure occurs in tension.

Above 2200°C the flexural strength decreases while the tensile strength continues to increase. It is believed that this effect is due to the mode of failure, which is no longer controlled by tension on the outer fiber, but rather by compression on the inner fiber. In this temperature range the compressive strength has already decreased below its room-temperature level, indicating that this mode of failure is probable (27).

The effect of the microstructure on the flexural strength is shown in Fig. 19. As predicted (16,29), the continuously nucleated materials are stronger than those nucleated on the substrate. Additionally, as the cone size decreases, the strength increases.

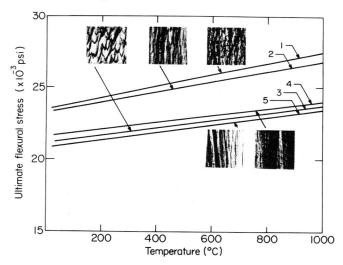

FIG. 19. Effect of microstructure on the flexural strength of pyrolytic graphite as a function of temperature (27).
Curves: 1, Fine-grained regenerative microstructure; 2, medium-grained regenerative microstructure; 3, coarse-grained regenerative microstructure; 4, medium-grained substrate-nucleated microstructure; 5, coarse-grained substrate-nucleated microstructure.

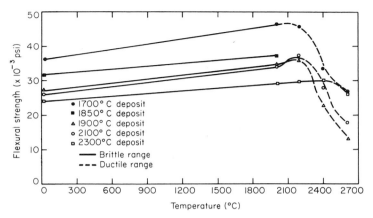

Fig. 20. Effect of deposition on flexure strength of pyrolytic graphite as a function of temperature (27).

The effect of deposition temperature on the flexure strength as a function of test temperature is shown in Fig. 20. An equivalency of density (2.18 to 2.20 gm/cc) and microstructure exists. Apparently a decrease in deposition temperature results in an increase in strength [50% higher at 1700°C (3100°F) than 2300°C (4172°F)].

Table V shows the effect of surface finish on the flexural and tensile

TABLE V

Effect of Surface Finish on the Strength of Regenerative Microstructural Pyrolytic Graphite

(2100°C Deposit)

| Direction | Surface finish, microinches | | Strength | | Standard deviation | Number of tests |
	In plane	Across plane	Tensile ($\times 10^{-3}$ psi)	Flexure ($\times 10^{-2}$ psi)		
A-B	35	35[a]	16.0		±1.20	7
A-B	35	12[a]	17.7		±0.90	6
A-B	35	6	18.7		±1.20	7
A-B	12	6	19.8		±1.00	5
C		30–60		18.8		10
C		12		20.7		10
C		1		21.9		10
C		32	0.730		±0.080	10
C		8	1.080		±0.030	10

[a] As machined.

strengths of continuously nucleated pyrolytic graphite. Very little difference is observed between as-machined and polished specimens. However, a 20% variation is observed when the polished surface is roughened. Apparently, also, surface finish is more critical in the C direction than in the AB direction, although this may or may not hold at elevated temperatures. As much as a 50% variation in C-direction tensile strength can occur when the finish is altered from 35 to 8 microinches. The C-direction flexure results show a less sensitive response.

5. Shear Strength

Two shear strengths can be defined for pyrolytic graphite, translaminar and interlaminar. Since pyrolytic graphite is an anisotropic material, experiments in which translaminar shear strength is measured are difficult to perform on sound material; therefore, few data are available (27,30). Standard torsion tests cannot be performed on plate material in a translaminar direction without interlaminar failure. Single and double direct-shear methods are complicated by plane-alignment problems. The limited usefulness of translaminar shear strength in design restricts interest to an academic basis. As determined by the double shear technique, the approximate value of this strength is 12,300 psi.

In contrast, the design-important interlaminar shear strength is shown in Table VI (17,27,30). Calculations show that a 3° misalignment of planes in a compression member will cause a premature interlaminar failure. The methods used to measure interlaminar shear are (a) torsion, (b) direct shear, (c) beam, and (d) compression. Data on these are compared in Table VI.

Gebhardt and Berry (30), using the torsional method, reported 1500 and 2800 psi for substrate-nucleated and continuously nucleated pyrolytic graphite, respectively. This corresponds with the 1350 psi measured by double direct-shear measurements for substrate-nucleated material. Single direct-shear measurements yield results that are approximately 50% lower than the double shear and torsional tests. The shear strength of continuously nucleated material determined by the beam method of testing was 2300 psi (good correlation with the torsional method).

The shear strength of the same type of material determined by the compression specimen method was 3200 psi. Although agreement among workers using this technique is good, frictional restraints at the loading surface usually give rise to comparatively higher results. It is felt that the surface stresses are sufficient to alter the principal stress direction to such a degree that the resultant interlaminar shear stress is lower than that calculated.

Only a limited amount of interlaminar shear testing has been done

TABLE VI
INTERLAMINAR SHEAR STRENGTH OF PYROLYTIC GRAPHITE

Method of measurement	Shear strength ($\times 10^{-3}$ psi)	Schematic representation	Reference
Torsion	1.5–2.6		22
Direct shear Single	0.6		5
Double	1.4		4
Beam	2.3		4
Compression	3.2		4

above room temperature. These data, although widely scattered, show no drastic change in the range from room temperature to a testing temperature of 2760°C (*31*). For translaminar shear, no data have been reported above room temperature.

6. Creep

Very little creep data have been obtained for pyrolytic graphite. Those reported have been described as preliminary (*19*). Unispecimen and multispecimen creep testing techniques have been used (*31*) with the results given in Figs. 21 and 22 and in Table VII.

TABLE VII

ACTIVATION ENERGY FROM UNISPECIMEN CREEP DATA FOR PYROLYTIC GRAPHITE

Temperature change, C°		Activation energy (kcal/mole)
From	To	
2480	2540	86
2540	2590	99
2590	2700	78
2700	2760	61
2760	2800	52

Note: Stress 14,400 psi parallel to basal planes.

A curve of elongation against time is given in Fig. 21. One way of obtaining activation energies from such a test is to compare the slope of the curve just prior to a temperature change with the slope just after such a change. Values of activation energy thus determined are given in Table VII. A second, but less desirable, method of obtaining activation energy from a unispecimen creep test is to obtain creep rate values from the slope of a straight line fitted to the last portion of the curve (Fig. 21) at each temperature level, and to plot these against the reciprocal temperature. Such a plot and the slope of the least squares straight line gives an activation energy of 114 kcal/mole.

In Fig. 22 we have a series of curves of elongation against time for various multispecimen creep tests. The problems of defining a meaningful creep rate for such tests have been discussed (32). For this study the slope of the curves at two values of elongation (5 and 10%) was taken. A plot of the creep rates thus obtained against the reciprocal tem-

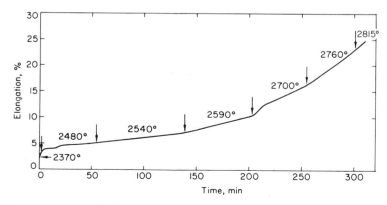

FIG. 21. Curve of elongation against time by unispecimen technique; stress 14,400 psi parallel to basal planes (19).

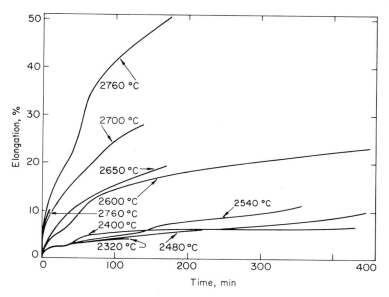

Fig. 22. Curve of elongation against time; stress 21,000 psi parallel to basal planes (19).

perature with a least-squares straight line fitted to the points, gave an activation energy of approximately 100 kcal/mole. The agreement with the other data (Table VII) and that reported for pitch-coke graphite (32,33) is considered to be good.

None of the equations that have been used to describe the creep behavior of pitch-coke graphites at high temperatures adequately describes the creep behavior of pyrolytic graphite. The most reasonable agreement was obtained by the parabolic relationship $C = A + bt^{-n}$, where A and b are constants, C is the creep strain, and t is the time.

The major difficulty in making such an analysis is due to the large data scatter, which should be reduced as more testing is done and as better quality material is available.

7. Fatigue

The 10^6-cycle fatigue-endurance strength and one-point-loading bend strength of 2540°C(4600°F)-annealed substrate-nucleated (2160°C deposition) pyrolytic graphite at room temperature has been measured (34). A mean value of 6500 psi with a ±2000 psi scatter band was obtained. From the 26 test results in the range of 4×10^5 to 3×10^6 cycles, the 10^6-cycle endurance strength for this particular pyrolytic sample is well established.

The limited data at 10^3 cycles (Fig. 23) indicate a sharp dropping

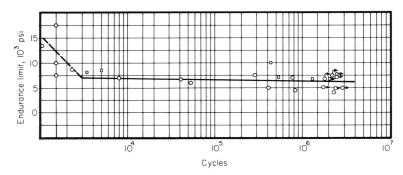

FIG. 23. Pyrolytic graphite fatigue S/N curve (34) Tests were made on cylindrical specimens of 2.24 density at room temperature. These were ground flat on the side of the last deposited layer. Cantilever-held specimens were tested at 1800 cycles per minute. ○, Virgin failure; □, retest failure; →, runout.

knee in the strength level between 1000 and 3000 cycles. The average amplitude for all tests was 0.110 to 0.120 inches or 0.055 to 0.60 inches in either direction. The strength at 1000 cycles was approximately 15,000 psi and dropped to approximately 7500 psi at 3000 cycles.

Test results showed that at room-temperature, the 10^3-cycle fatigue strength and the bend strength are approximately equal: 15,000–16,000 psi.

No failure mechanism was established conclusively. Visual observations indicated that delamination occurred at some point before failure, but not as a cycle-dependent function; specimens that endured 10^6 cycles did not show delamination, yet when brought to failure, all specimens showed delaminations (34).

8. Hardness

The hardness of pyrolytic graphite at room temperature has been measured in both the AB and C directions (35). The results are summarized in Table VIII for material of 2.20 gm/cc density.

TABLE VIII
HARDNESS IN THE AB AND C DIRECTIONS OF PYROLYTIC GRAPHITE[a]

	Measurement method		
Direction	Knoop, kg/mm²	Rockwell "L," 60-kg load	Shore
AB	27–28	Split	70–75
C	84–91	127	100–105

[a] Density 2.20 gm/cc.

B. Thermal Properties

1. Thermal Conductivity

The thermal conductivity of pyrolytic graphite in the AB direction is shown in Fig. 24. Although data for temperatures higher than 2700°C (4900°F) are generally lacking, some work on graphite suggests a drastic decrease in conductivity above 3200°C (5800°F) (37). This effect has been explained as caused by scattering from thermally activated vacancies (20). The conductivity curve is reversed at low temperatures, giving an exponential temperature dependence of 2.5 from 10° to 80°K (38,39). Thus, a peak of 3300 Btu-in per hour-square foot-°F is reached at about room temperature. In the AB direction, consequently, pyrolytic graphite is one of the very best heat conductors among elementary materials.

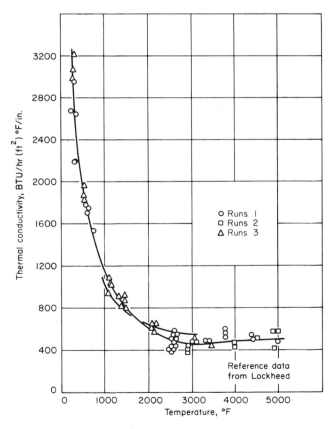

Fig. 24. Thermal conductivity of pyrolytic graphite in AB direction (36).

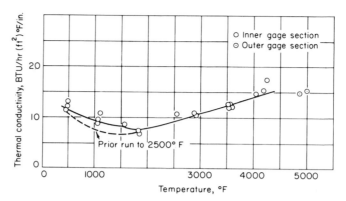

FIG. 25. Thermal conductivity of pyrolytic graphite in C direction (precut strip 3.0 × 0.5 × 0.25 inches) (36).

Figure 25 shows thermal conductivity in the C direction. Interestingly, pyrolytic graphite is particularly anisotropic with respect to thermal conductivity, values at room temperature being about 300 times as high in the AB direction as in the C direction. As indicated in Fig. 25, a gradual increase in conductivity begins at about 1200°C (2200°F). This increase has not been satisfactorily explained, although the suggestion has been made that microcracks or open pores might be responsible because they permit heat transfer by radiation (20). Since the amount of crack and pore area that would be required to accomplish this is much greater than that observed, it is likely that sufficient authoritative tests have not been made to prove conclusively that the increase occurs. If a prior run is made to 1370°C (2500°F), as shown in Fig. 25, the interlaminar distance is spread and the thermal conductivity is lowered moderately in the 260°–1000°C range.

The conductivity peak in the C direction is reached between 100° and 200°K, and then falls as T^{-1}. There is an exponential temperature dependence of 2.3 in the conductivity curve at low temperatures (38,39).

The thermal conductivity is quite sensitive to both heat treatment and deposition temperature. A 3-hour heat treatment at 2550°C has been shown to increase the C-axis conductivity by about 50%. Decreasing the deposition temperature lowers conductivity in both the AB and C directions. Also, a highly regenerative graphite has been noted to have a higher thermal conductivity in both directions as compared with a surface-nucleated material.

2. Thermal Expansion

The thermal expansion of pyrolytic graphite is markedly anisotropic. Up to 1000°C the total expansion in the C direction is positive. The AB-direction thermal expansion, however, is negative at room temperature, reaching a point of inversion between 150° and 400°C. At higher temperatures it becomes positive. Heat treatment reduces the thermal expansion in the AB direction but increases the coefficient in the C direction. The thermal expansion coefficient of annealed graphite is similar to that reported for single crystals (36). At temperatures above 2600°C pyrolytic graphite undergoes an irreversible dimensional change, and thermal expansion curves taken in excess of this value generally exhibit a hysteresis effect. Figure 26 shows thermal expansion data to 2760°C (5000°F) for both orientations. Continuously nucleated pyrolytic graphite shows a higher expansion in the AB direction than surface-nucleated material. The C-direction value is not greatly influenced. This condition is probably related to orientation effects (36). There is evidence that the expansion coefficient varies through the thickness of the deposit (16,36,41).

3. Specific Heat

At low temperatures the behavior of pyrolytic graphite can be expected to resemble that of single-crystal graphite. At elevated temperatures, all graphite shows the same behavior. Figure 27 gives specific

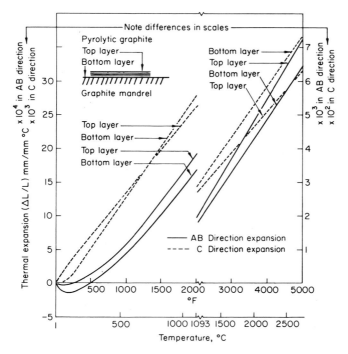

FIG. 26. Linear thermal expansion data to 2800°C (5000°F) (40).

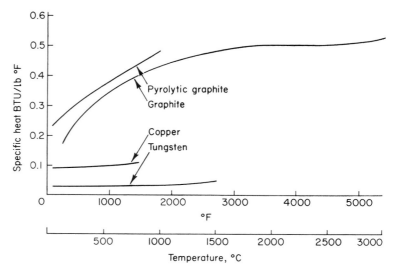

Fig. 27. Curve of comparative specific heat against temperature (37).

heat values for several graphites and for two familiar metals for comparison. At low temperatures the specific heat varies linearly with temperature, reaching a value of about 0.16 Btu/lb-°F at room temperature (42). At about 400°C (673°K) the curve begins to depart significantly from the DuLong and Petit value of 0.5 Btu/lb-°F at about 1500°C (1773°K). Such behavior is characteristic of monatomic solids, and is theoretically well understood. The behavior of the specific heat above 3300°C (3573°K), however, is quite anomalous. The sudden rise in specific heat has been attributed to the thermal creation of vacancies (37). Recent electron microscope studies of vacancy-controlled processes in graphite, however, make this interpretation uncertain (43).

Specific-heat data on pyrolytic graphite taken by means of a calorimetric technique show good agreement with the theoretically derived values.

4. Emissivity

Pyrolytic graphite is anisotropic with respect to emissivity, as would be expected from the anisotropy shown in electric resistivity measurements. In the AB plane, the electrical conductivity (that is, the electron mobility) is relatively high. Therefore, the reflectance is higher in the direction normal to this plane and by Kirchhoff's law, the emissivity is lower. In a plane perpendicular to the AB plane, the situation is more complicated. The resistivity in the C direction is higher, resulting in a high emissivity for that component of light with an electric vector vibrating in the C direction. The component whose vector is perpendicular to the C direction has a lower emissivity. Light emitted from a plane perpendicular to the AB plane should, therefore, be strongly polarized in the C direction and should be characteristic of a higher emissivity. The spectral emissivity has been reported as a function of wavelength, both when the layer planes are perpendicular to the plane of emissivity and when they are parallel to the plane of emissivity (44). Although considerable spread in data is observed, the AB plane values are lower than the C-plane values. The spectral emissivity does not appear to change with temperature. At a wavelength of 0.65μ, which is used for most pyrometric measurements, a value of 0.78 has been obtained for the normal spectral emissivity in the AB plane.

The total normal emissivity has been obtained by integrating the spectral emissivity over the wavelength region of 0.20 and 0.35μ. A value of 0.53 was obtained at 1000°C for an as-deposited surface having a 200–300μ surface finish. Normal emissivity of pyrolytic graphite as a function of temperature is shown in Fig. 28. No total hemispherical emissivity data have been reported from direct measurements, but calculated results from use of a calorimetric technique are shown in Fig. 29. In these studies thin layers of pyrolytic graphite were deposited on graphite

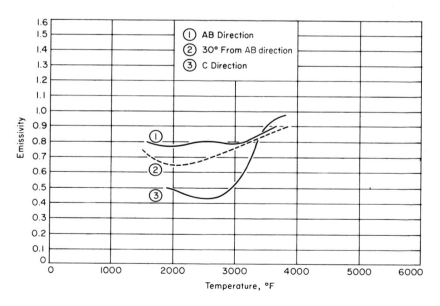

FIG. 28. Total normal emissivity of pyrolytic graphite as a function of temperature (tests by Southern Research Inst.).

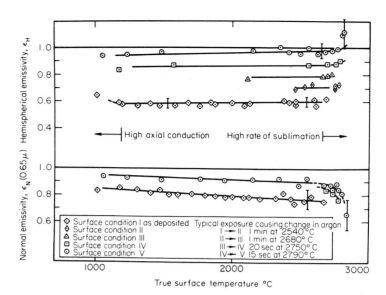

FIG. 29. Total hemispherical emissivity and normal emissivity of pyrolytic graphite samples at 0.65 μ (44).

cylinders (45). The total emissivity appears to be about 0.6 at 1000°C with a slight positive temperature slope. The C-plane value appears to be about 0.9.

Emissivity is quite sensitive to surface finish, and polishing the surface of most materials, including graphite, generally lowers the emissivity. This is not true of pyrolytic graphite. Some investigators have found that polishing increased both the normal spectral and total normal emissivity (44). Apparently, the act of polishing converted the highly oriented AB surface to a surface containing both AB and C surfaces. Polarization effects have also been noted with pyrolytic graphite (46). There are some indications that Lambert's law is not obeyed.

5. Optical Constants

The authors are aware of no data published on the optical properties of pyrolytic graphite. Owing to its highly oriented nature, these properties should be similar to those of graphite single crystals. The index of refraction of pyrolytic graphite varies from 2.15 with the electric vector of the light wave in the layer plane, to 2.00 with the electric vector across the layer planes. The absorption index varies from 1.42 with the electric vector parallel to the layer planes to 0.02 with the electric vector perpendicular. Reflectance and polarization of light reflected from a graphite crystal at a 45° angle of incidence, and reflection with layer planes parallel and perpendicular to the surface, have been theoretically approximated using extension of the isotropic metallic reflection formulas; theory is in good agreement with experimental data. Pyrolytic graphite has been used in thin sections as a transmission polarizer for infrared radiation.

C. ELECTRICAL AND MAGNETIC PROPERTIES

1. Electrical Resistivity

Pyrolytic graphite is highly anisotropic with respect to electrical resistivity. Room-temperature values for the resistivity in the AB plane are about 700 microhm-cm. Across this plane, the C-direction resistivity is about 500,000 microhm-cm. Thus, in the AB plane, pyrolytic graphite is a good conductor (Fig. 30); whereas in the C plane it is a poor conductor. This has given rise to a number of novel uses for pyrolytic graphite, particularly for high-temperature resistance furnaces. Although often attributed to the high degree of orientation of crystallites, the anisotropy in the electrical resistivity is considered to be caused by the presence of microcracks and other gross microstructural defects (49).

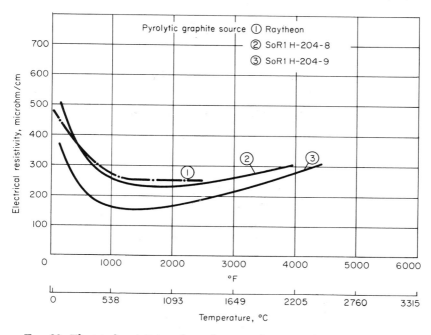

FIG. 30. Electrical resistivity of pyrolytic graphite as a function of temperature, AB direction (Lockheed).

The variation of the electrical resistivity in the AB plane with annealing temperatures is shown in Fig. 31. The spread represents the scatter observed in testing a large number of samples machined from a large block. At room temperature, the resistivity has a negative temperature coefficient as shown in Fig. 30. At about 1000°C (1830°F) it reaches the lowest value and then begins to increase. The electric resistivity in the C direction shows no such inflection point, decreasing steadily with temperature.

The electrical resistivity has been found by Wagoner and Eckstein to depend strongly on the deposition temperature (50). The data of Klein

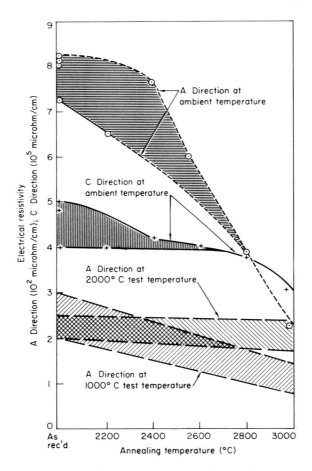

Fig. 31. Effect of annealing temperature on electrical resistivity of pyrolytic graphite (41).

are shown in Table IX. Klein (47,49,51) has found that the electrical resistivity as well as the other electron transport properties are dramatically changed by heat treatment. These data are shown in Table X.

Heat treatment at 3600°C causes the temperature coefficient of resistivity in the AB direction to decrease and approach that of a single crystal. The effect of annealing on the C-direction resistivity is also to lower the absolute value considerably. This behavior is shown in Fig. 31 (41).

TABLE IX
Electrical Characteristics of As-Deposited Pyrolytic Graphite
(Room Temperature)

Deposition temperature, °C[a]	Galvanomagnetic parameters		Resistivity parameters	
	$\Delta\rho/(\rho_0 H^2)$, $10^{-10}\ G^{-2}$ [b]	R_0, cm^3/°C[c]	ρ_{ab}, $10^{-4}\ \Omega$-cm	ρ_c, Ω-cm
$\lesssim$1900	−0.035	+0.195	23.5	0.072
2100	+2.1	+0.33	5.65	0.49
2300	+3.8	+0.31	4.70	0.62
2500	16.0	+0.099	3.50	1.20

[a] Presumably correct within ±50°C, except for the $\lesssim$1900°C specimen.
[b] As determined from a field-dependence plot of the magnetoresistance at 293°K.
[c] As determined from a temperature-dependence plot of the Hall coefficient at 2520°C.

The effect of pressure up to several hundred kilobars on the resistivity of pyrolytic graphite has been measured (52). The AB-direction resistivity was found to increase at higher pressure, apparently due to recrystallization. The C-direction resistivity, on the other hand, was found to decrease with increasing pressure and level off at high pressure, apparently due to strongly decreasing compressibility.

2. Magnetic Susceptibility

The magnetic susceptibility of pyrolytic graphite is much larger than that of any other form of graphite (50). The trace susceptibility, which is the sum of the susceptibilities in three orthogonal directions, is often used to compare polycrystalline specimens with single-crystalline material. It does not depend on the degree of preferred orientation of the crystallites. The trace susceptibility of pyrolytic graphite is 33.5 × 10^{-6} emu/gm, which is to be compared with 22.8 × 10^{-6} emu/gm for the best single-crystal material. The results of individual measurements are shown in Table XI. The susceptibility in the C direction is seen to be much greater than that in the other two directions. This anomalously large trace susceptibility is attributed to turbostratic stacking. Two-dimensional

TABLE X

ELECTRICAL CHARACTERISTICS OF HEAT-TREATED PYROLYTIC GRAPHITE (BASAL-PLANE PROPERTIES)

Heat treatment[a] °C	Resistivity, 10^{-6} Ω-cm			Hall coefficient,[b] cm³/°C			Magnetoresistance,[b] %		
	20.5°K	77°K	293°K	20.5°K	77°K	293°K	20.5°K	77°K	293°K
None	840	780	450	+2.90	+1.80	+0.32	−1.7	−0.53	0.19
2500	530	420	240	−0.14	−0.099	−0.046	0.65	0.64	0.33
2750	96	126	110	−0.39	−0.16	-0.066_5	19	7.1	1.5
3000	35	48	50	−0.68	−0.42	-0.090_5	260	64	7.1
≥3500	6.1	24	42				8800	250	12

[a] For 1 hour at the indicated temperature.

[b] As measured on applying a transverse magnetic field of 2520 G.

TABLE XI

RELATION OF DIAMAGNETIC SUSCEPTIBILITY OF PYROLYTIC GRAPHITE TO
DEPOSITION TEMPERATURE

Deposition temperature, °C	Diamagnetic susceptibility, 10^{-6} emu/gm			
	$X\perp$ (AB direction)	$X\perp$ (BA direction)	$X\|$	X Total
1600	-2.77	-2.81	-7.71	-13.2_9
1800	-3.88	-3.88	-16.1_8	-23.9_4
2000	-2.87	-2.88	-21.7_7	-27.5_2
2200	-3.18	-2.91	-25.4_1	-31.5_0
2300	-2.94	-2.98	-27.0_0	-32.9_2
2400	-3.74	-3.79	-26.0_0	-33.5_3
2500	-3.10	-3.30	-21.1_0	-27.5_0

graphite, which has no interlayer interaction, has a theoretical trace susceptibility of 39×10^{-6} emu/gm, and pure hexagonal graphite, where interlayer interactions cause a band overlap, has a value of 22.8×10^{-6} emu/gm. On theoretical grounds, McClure and Yafet have estimated the diamagnetic susceptibility of fully turbostratic graphite to be 80% as compared with the observed value of 86% for pyrolytic graphite (53).

3. Hall Coefficient and Magnetoresistivity

The electron transport properties of pyrolytic graphite have been extensively studied by Klein and various co-workers (47,49,51). The absence of binder materials in pyrolytic graphite, as well as its near-theoretical density, has generated considerable interest in its electron transport properties. Klein found that the resistivity, Hall coefficient, and magnetoresistivity can be adequately described in terms of an energy-band model and a suitable scattering law. Room-temperature values for these properties are shown in Table IX as functions of deposition temperature. The properties are seen to be quite sensitive to these processing parameters.

Klein has also shown that heat treatment dramatically changes the directional electron transport properties (Table X). The Hall coefficient of as-deposited pyrolytic graphite is positive, typical of a p-type semiconductor (47). Upon heat treatment of 2500°C, the material reverts to n-type. The dramatic increase in magnetoresistivity has been attributed to improvement in crystallite alignment with increasing temperature. The Hall effect has been found to have a negative temperature coefficient which decreases to zero at about 500°C, where it becomes positive.

D. CHEMICAL PROPERTIES

1. Purity

The purity of pyrolytic graphite is governed primarily by the purity of the feed gases, the substrate, and the other graphite materials used in the hot zone. The use of CP-grade raw materials can greatly improve the purity of the product. A typical spectroscopic analysis shows the principal impurity to be silicon at 0.001%, and calcium, aluminum, magnesium, nickel, iron, copper, and titanium at levels less than 0.0001%. The ash content usually ranges between 0.001 and 0.004% by weight. A typical spectroscopic analysis of the ash content is given in Table XII.

TABLE XII
COMPOSITION OF ASH FROM PYROLYTIC GRAPHITE

Element	Percentage	Element	Percentage
Si	50	Ti	0.7
Al	6.1	Mn	0.1
Ca	5.0	Cu	0.03
Zr	3.3	Mg	0.003–0.03
V	2.7	Ta, W	1.0
Fe	2.6	Ni	0.5
Cb	1.6	B, Ce	0.03

A statistical study comprising 204 tests was made of pyrolytic graphite produced in 26 different furnace runs. Total ash content was 0.0035% with a standard deviation of 0.0027%. The material was produced from natural gas as the source of carbon and with conventional graphite materials not specially purified.

2. Oxidation Resistance

Graphite oxidizes appreciably at temperatures above 450°C. The carbon-oxygen reaction is still not completely understood although it has been studied extensively (54). However, there is general agreement that at these lower temperatures, the rate of mass loss is determined by reaction kinetics. As the temperature rises, a transition occurs to a diffusion-controlled regime where the rate of mass loss is relatively independent of surface temperature. Above about 4700°F, the rate of sublimation exceeds the rate of oxidation.

Pyrolytic graphite is superior to normal graphites in the regime controlled by reaction rate (54). This is due to the absence of binder materials, which tend to oxidize more readily than the crystallites. In

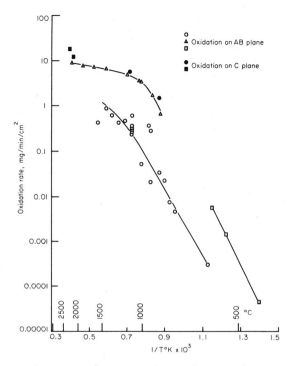

FIG. 32. Oxidation rate for pyrolytic graphite.

addition, the low porosity of the pyrolytic graphite decreases the effective surface area for oxidation.

Preferential oxidation occurs at the edge of crystallites. Hence, the degree of orientation of the material, as influenced by deposition rate or deposition temperature, might be expected to influence the oxidation behavior. The oxidation in air of pyrolytic graphite is shown in Fig. 32.

3. Chemical Compatibility

Pyrolytic graphite is strongly resistant to chemical attack. It is inert to water, organic solvents, and most acids. The exceptions are chromic acid and chlorate solutions, by which it is strongly attacked. Chemical attack in graphite seems to occur preferentially at the crystallite edges. Since relatively few crystallites are exposed in pyrolytic graphite, it is not surprising that it is generally better than commercial graphite in this regard. Attack by chromic acid solution has caused exfoliation and separation of layers and formation of black powder on the surface, but no change in the C-layer spacing. Attack by a chlorate solution has also caused exfoliation and separation of the layers, with a change in the C-layer spacing. Greater attack has been noted on machined surfaces than on the "as-deposited" surface.

E. NUCLEAR PROPERTIES

Radiation can produce permanent effects in graphitic materials by atom displacement and electron excitation. These effects have been extensively investigated in graphite. Of interest currently, however, is the radiation stability of pyrolytic graphite.

Dimensional changes observed after exposure of pyrolytic graphite to neutron radiation at 2640 and 4870 megawatt-days per atom were —0.3% in the AB direction and +0.75% in the C direction. The radiation produced a steady increase in the coefficient of expansion in the C direction, but essentially no net change in the AB direction. The initial values were $\alpha_c = 20.5 \pm 0.1$. After an exposure of 4870 Mwd $(At)^{-1}$, the following values were found, $\alpha_{ab} = 0.07 \pm 0.04$ and $\alpha_c = 24.6 \pm 1.0$. These dimensional changes were accompanied by changes in the crystal lattice parameters. After irradiation at 2640 Mwd $(At)^{-1}$ and 650°C, the a_0 spacing was found to increase by +0.11 ± 0.05%, whereas the AB direction increased by +0.5 ± 0.2%.

Since pyrolytic graphite is elemental, its nuclear properties are identical to those of normal graphite. Graphite is a commonly used reactor material and its nuclear properties have been extensively investigated. The average logarithmic change in energy per unit energy ϵ, is 0.1589. The absorption cross section, σ_a, at 2200 m/sec is 0.0032 barns. This value, however, is for elemental carbon, and the scattering cross section for engineering material is usually higher, depending on the amount and kinds of impurities present. The average cross section, σ_{s1}, for thermal neutrons is 4.8 barns. The epithermal cross section, σ_s (epi), which is the scattering cross section free of binding effects, is 4.66 barns. Since carbon is a typical $1/v$ scatterer, the scattering cross section at very high energies is negligible. The average angle through which the neutron is scattered is 0.05598 radians, based on a value of 11.908 for the mass of the carbon nucleus relative to the neutron mass. Their nuclear properties make the graphites excellent moderator materials since they are quite effective in slowing fast neutrons to thermal energies while absorbing few of the neutrons themselves.

V. Residual Stresses in Closed Shapes

As noted in the preceding section, there are large differences in both mechanical and thermal properties of graphite, depending on whether it is considered in the AB direction or the C direction. These include a 300 to 1 difference in thermal conductivity and a 14 to 1 difference in thermal expansion coefficients as examples (at a specific temperature). Even the modulus of elasticity, the strength, and the Poisson's ratio are significantly different in the two principal directions. This combination of properties results in the development of severe stresses in closed shapes, which can lead to premature failure or render parts impossible to manufacture. Residual stresses develop in all closed shapes (cylinders, cones, hemispheres, etc.) when they are cooled from 4000°F deposition temperature. A flat sheet of pyrolytic graphite can be heated and cooled without setting up these residual stresses since the part is free to move in all directions. To design with the material it is necessary to have a thorough understanding of where these residual stresses arise.

The residual stresses arise as follows: (a) cooling from the deposition temperature coupled with anisotropic thermal expansion, (b) variations in the microstructure of the deposit caused by process variables, i.e., whether the material is substrate nucleated or continuously nucleated, (c) secondary anisotropic expansion effects due to deposition surface roughness on a microscopic scale, (d) temperature and thickness gradients within the deposits and (e) effects related to female vs. male mandrels and geometrical discontinuities. Considerable work has been done on analyses of residual stresses in pyrolytic graphite (23,55–58).

By far the largest factor in determining residual stresses in a closed shape is the stress set up on cooling from the deposition temperature, due to thermal expansion anisotropy. The circumferential tensile stress produced at the inner surface of a closed shell is given by the formula

$$\sigma = \frac{E_a(\alpha_c - \alpha_a)\Delta T}{2(1 - \mu_a^2)} \frac{t}{\bar{R}}$$

where the E_a and μ_a are the elastic modulus and Poisson's ratio in the AB or layer direction; α_c and α_a are the thermal-expansion coefficients in the C and AB directions, respectively; ΔT is the temperature change; t the thickness of the shell and $\bar{R}$ the average radius. The ratio of $t/\bar{R}$ is often used to judge whether or not a part can be successfully produced. It has been found from practical experience that in most cases if the ratio $t/\bar{R}$ is less than 0.07, the part can be manufactured without too much difficulty. Parts have also been made when this ratio was as high as 0.15, but in general such parts may contain delaminations or cracks. A plot of the residual tensile stress on the inside of a cylinder at various $t/\bar{R}$ ratios is given in Fig. 33. For a typical $t/\bar{R}$ ratio of 0.04, the calculated tensile stress is 4300 psi. Since pyrolytic graphite has a tensile strength of 10,000 psi at room temperature, the part would not be ex-

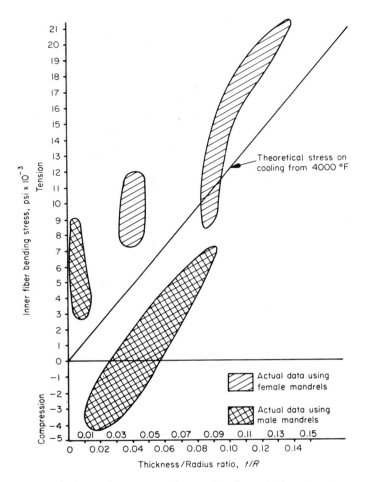

FIG. 33. Residual tensile stress on the inside of a cylinder at various ratios of thickness to radius.

pected to fail. Unfortunately, nodules or other growth abnormalities can cause stress concentrations and markedly reduce the usable working stress so that, for design considerations, a factor of safety of 2 is usually used. As can be seen from Fig. 33, the actual stress measured on parts is quite different from that predicted by theory. It has been found that for deposits made on male mandrels, i.e., where the deposits are on the outside of the mandrel, the measured residual stresses are considerably higher than predicted. For female mandrels, the reverse is true. Coffin, Levy (23,55), and others have shown that the discrepancy is a result of stresses from the growth of the deposit during deposition, and secondary stresses due to the surface roughness of the substrate. In general, these secondary stresses are compressive on the inner surface of female mandrels and subtractive (reducing stress levels), whereas they are tensile and additive on the inner surface of male mandrels. Parts produced on female mandrels, therefore, can be made with much higher $t/\bar{R}$ ratios than those for parts made on male mandrels. When they fail, parts made on female mandrels generally fail in tension on the inside surface by cracking normal to the deposition planes, while parts made on the male mandrels fail by delamination between the layer planes.

Another major source of internal stress is geometrical discontinuities (edges, changes in thickness, holes, etc.). Great care must be exercised in the design of mandrels and in the machining of pyrolytic graphite parts so as to minimize these stresses. In machining of closed shapes it is desirable to take small cuts on the inside, then on the outside surface, so as to prevent an unbalancing of stresses. Nodules of a large size can be avoided by using fine-grained graphites as mandrels, by cleaning them free of dust particles, and by developing a continuously nucleated microstructure.

To date, no one has found a satisfactory method for removing these residual stresses either during the deposition process or by some post-deposition heat treatment. Many studies have been undertaken and some limited success has been achieved by special contouring of the deposition mandrel (23). However, this technique has led to other undesirable effects. Attempts to remove the stresses by very high temperature annealing have been predestined to failure by the generation of larger ΔT's and, therefore, greater anisotropically generated stress. Other measures to reduce the stresses by establishing large thermal gradients across the thickness during the deposition process have also failed. To date, the "magic number" for $t/\bar{R}$ ratio for consistent delamination-free and crack-free parts remains at 0.07.

VI. Inspection and Quality Control

In 1959, not a single written specification existed for pyrolytic graphite. Today all major producers and users of pyrolytic graphite have published specifications generally acceptable to all. The inspection and quality control procedures developed are not significantly different from those used for ceramics or the more brittle transition elements. The major areas where innovation has been required have been in extremely high temperature mechanical testing (above 3500°F) and in all phases of thermal and electrical property determination.

In an effort to improve the quality-control techniques as one aspect of insuring improved reliability, a Committee, C-6 on Pyrolytic Materials, was established by the American Society for Testing and Materials early in 1963. A subcommittee on Material Specifications and Test Methods was formed to establish industry-wide standards. From the activities of this group have come specifications that cover most of the requirements in presently accepted commercial practice.

Test methods have been well established for many conditions, and the data can be used in engineering calculations with the same degree of confidence enjoyed in the use of more common brittle engineering materials. There is a rather large standard deviation for mechanical properties, even at room temperature.

Careful sample preparation and inspection is necessary to assure reliable and reproducible results. As with most brittle materials, pyrolytic graphite has a high degree of notch sensitivity up to temperatures of 4500°F, and great care must be taken in preparing samples for mechanical tests.

The standard deviation associated with thermal property data obtained for establishing specifications is very small. Analysis of these data for run-to-run variations, position effects, or other forms of bias such as sample preparation indicated no trends beyond the normal Gaussian distribution observed in unbiased data. Fortunately, enough data were available to permit statistical treatment of the information.

Although property measurements and their specifications are an important factor in quality control, this is not the entire story. Certain types of defects can exist in pyrolytic graphite which are detrimental to its performance from both a thermal and mechanical standpoint. These defects are nodules, delaminations, cracks, and bow. All of these are readily visible and easily specified as long as they intersect free surfaces. Nodules and bow are always plainly and painfully visible. Most manufacturers and users of pyrolytic graphite specify allowable nodule size and distribution, and bow tolerance in the as-deposited parts. The formation of nodules has been covered in great detail by Coffin (23) and their effects on mechanical properties by a number of authors (18,56,59).

Bow in the as-deposited plate can significantly affect the thermal performance of parts. Much of the insulating value of pyrolytic graphite can be lost if parts are cut either from badly bowed plate or cut at an angle to the deposition surface. Certain users of pyrolytic plate now specify the maximum allowable bow in as-deposited plate, and also how the material must be cut in reference to the first deposited surface. It appears almost impossible to fabricate large plates of pyrolytic graphite without some curvature in the as-manufactured condition.

Cracks and delaminations that intersect the surfaces are readily discernible by conventional dye penetrant or other standard techniques. If, however, the delaminations or cracks do not intersect a free surface, some difficulty may be encountered in finding them by nondestructive methods. At the present time, ultrasonic inspection appears to be capable of finding most delaminations inside plate. The procedure is similar to that used on metals and few modifications are necessary. Studies are being conducted currently to better correlate the size of the defect with the reading of the instruments.

X-ray radiographic procedures for finding internal delaminations or cracks have proven somewhat less reliable than the ultrasonic method. With copper radiation and relatively low voltage, radiographs have been produced which indicate internal delaminations. Unfortunately, graphite is quite transparent to X-ray radiation and the interpretation of the photographs is sometimes inconclusive.

Although much of the detailed deposition process is considered proprietary by the various producers of pyrolytic graphite, there is some acceptance of process control as a means of quality control today and this will probably increase in the near future. The many excellent specifications obtained from manufacturers and users give further details on exact inspection and quality control procedures.

VII. Applications

A. Nose Cones and Reentry Heating

The closer a reentry vehicle comes to earth, the greater the density of air it must tunnel through to land. As mechanical work is done on the air, the air becomes heated in the vicinity of the reentry body. The energy going into heating the air is proportional to the vehicle's decrease in velocity (kinetic energy). At the leading edge or tip of the protruding surface a stagnation region exists, and heat flux is highest at this point.

It is possible to design reentry vehicles so that the heat flux is spread over a large area. Examples of vehicles so designed are Mercury, Gemini, and Apollo. Slender vehicles (whose form approaches a sharp stagnation point) concentrate the energy transfer on a small area; therefore, very high heat fluxes are generated. Examples of such vehicles include the ballistic missiles. To decide what materials to choose in constructing the reentry vehicle, a trajectory-mission analysis must be performed. The analyst determines the pressure and enthalpy and, judging by past observations of material behavior under such test environments, chooses a suitable material. Fortunately, considerable experience has been gained with materials exposed to reentry environments. Information has been gathered under such variable conditions, however, that it cannot be presented in a concise manner to show true comparisons between pyrolytic graphite and other materials.

The question arises, why should graphite be used for a nose-tip and heat-shield material? Carbon is the most refractory element known in that it possesses the highest melting point and strength at very high temperatures. The strength of graphite, comparatively low at room temperature, increases and becomes significant at high temperatures. For example, recent advances in reinforced pyrolytic graphites have produced materials with strengths of 6000–8000 psi to 15,000–20,000 psi at 2200°C (4000°F). This development is of great interest, since heat shield materials for reentry vehicles are subjected to high shear loads in the 3000°C (5000°–7000°F) temperature range. Vaporization and mechanical removal mechanisms also are operative, leading to the natural choice of graphite to resist these actions. Thus, the anisotropic physical properties of pyrolytic graphite have led to investigations of its usefulness for these demanding reentry applications.

In the late 1950's the Special Projects Office of the Navy Department initiated a program to develop a nose tip and heat shield for the Polaris missile from pyrolytic graphite. Unfortunately, the need arose in advance of industrial capabilities to produce such parts, and the tip and heat shield were eventually made from other materials. However, the program supported research and development so that in the early 1960's vehicles could be produced with noses having smaller radii. Parts made from pyrolytic graphite survived in applications where other temperature-ablative materials proved unsatisfactory.

The mass loss of carbon/graphite at elevated temperatures is shown in Fig. 34. The value is nondimensionalized, it is defined by comparison with the mass flow of air. The range of surface temperature between room temperature and a maximum of 2000°K (3140°F) can be considered the *oxidation regime* of up to 1 atm pressure. During this interval oxygen from moving air diffuses to the carbon/graphite surface, reacts to form an oxide of carbon, and leaves the surface as a gas.

Above 2000°K (3140°F) up to about 3000°K (4940°F) or perhaps slightly before (reportedly above 4750°F), there exists the *diffusion-limited regime*. Here, the carbon/graphite surface is oxidizing and yielding gaseous products at such a rate that larger amounts of oxygen from fresh air can no longer reach the surface and continue to increase the oxidation rate. Thus, the mass loss reaches a plateau where increased temperature (to 3000°K) no longer effects an increase in mass loss.

However, above 3000°K (4940°F), carbon atoms are beginning to shear off at a rate that increases with increased pressure, and the mass loss soars asymptotically. Here, we begin to understand something of

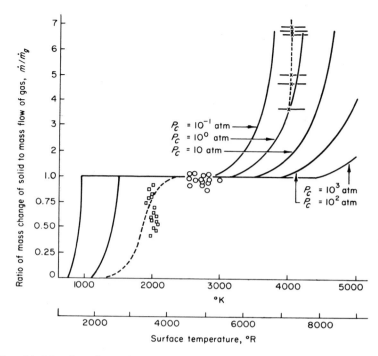

FIG. 34. Mass-loss data of pyrolytic graphite. ——— Theoretical curve [Scala and Gilbert (60)]; $\times P_e \sim 1$ atm; $\bigcirc P_e \sim 0.056$ atm; $\square P_e \sim 0.00825$ atm; ——— experimental data curve.

the problem of sharp-nosed carbon/graphite vehicles on reentering the earth's atmosphere at severe entry angles. Inches of material can be lost in a few seconds. It is obviously meaningless to talk of protective coatings in this *vaporization-controlled regime*. Even intimate mixing of other elements and carbides with carbon/graphites in this regime only serves to lower the vaporization temperature and thereby increase mass loss. It should be stressed that these measures must be restricted to the oxidation-controlled regime.

The major problem encountered in constructing nose tips and heat shields of pyrolytic graphite is in manufacturing sharp-radius parts of sufficient thickness, uncracked and undelaminated. As discussed previously, it is difficult if not impossible to fabricate parts for which the thickness-to-radius ratio is much greater than 0.1. Therefore, a nose tip of 0.25 inch radius is limited in thickness to 0.025 inch. Under most hypersonic reentry conditions this would not be of sufficient thickness to withstand oxidation and sublimation losses. Shear forces are also likely to be high, and as can be seen from the section on mechanical properties, pyrolytic graphites are very weak in shear between the planes in the AB direction. For these reasons, pyrolytic graphite will probably not find direct application in the nose tip area until these problems can be overcome. However, in areas removed from the nose tip, such as the heat shield, pyrolytic graphite holds significant potential. The size of defect-free parts that can be manufactured is somewhat limited; for example, the probable occurrence of an oversized nodule that would weaken the structure becomes high as the area of the part approaches several square feet. Furnace yields for such parts would be very low and costs very high. Considerable investigation is now going on to determine if pyrolytic graphite, because of its low specific gravity and extremely low thermal conductivity in the C direction, can be used as a substructure in reentry bodies to function as a thermal barrier. Here the structural integrity of the part would not be as critical a factor.

B. Thrust Chambers and Rocket Nozzle Throats

The use of pyrolytic graphite in thrust chambers and rocket nozzle throats today accounts for the largest volume use in the aerospace industry. This volume use began shortly after the commercial production of pyrolytic graphite in the late 1950's. Although the environmental conditions of rocketry are different from those of reentry vehicles, the two applications have in common extreme high temperatures and a relatively short operational period.

The initial application in this area was for nozzle throats and portions of the entrance and exit cones made into a single unit, with the pyrolytic graphite oriented to be an insulator in the radial direction. Such a chamber is shown in Fig. 35. Such a configuration, backed up with conventional graphite, plastics, steel, etc., can carry the main portion of the structural loads. When compared with chambers using other refractory materials, the chambers so prepared had great potential advantages of light weight, excellent insulation properties in the radial direction, good conductivity axially, and excellent resistance to thermal stress failure. However, initial experiments uncovered a number of serious problems. It was found to be extremely difficult to produce relatively large shapes free of defects (particularly oversize nodules). These nodules, acting as stress raisers, resulted in premature failure of the entire body.

Under oxidizing conditions, the area change in the throat was no better than with refractory metals, and, unless special care was taken with fuel-oxidizer mixing, the erosion might be considerably worse. Residual stress in the thick small-diameter deposit near the throat area often caused cracks, or the material there was so highly stressed that the

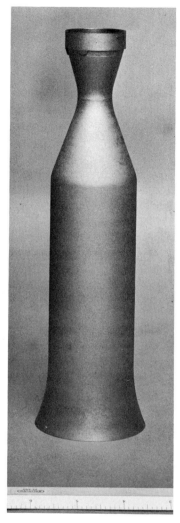

Fig. 35. Free-standing thrust chamber.

thermal and mechanical shocks of firing produced failures. These early failures led many to completely abandon any hope of using this material as a rocket nozzle throat.

With the advent of higher temperature solid rocket fuels containing large percentages of aluminum, the weight advantage inherent in the use of pyrolytic graphite, along with its good high-temperature mechanical properties, was more and more advantageous and could not be ignored. In what appeared to be a simultaneous development of a new nozzle concept, several companies, (General Electric, High-Temperature Materials, Curtiss-Wright, and others) evolved the washer or edge-grain nozzle design shown in Fig. 36. This nozzle worked on the principle of using a high-conductivity heat-sink material to keep the nozzle throat cool and thus prevent erosion by maintaining material strength. For example, it has been shown by two independent investigating teams that if the surface temperature of the pyrolytic graphite washers in contact with the hot gas stream is kept below 4500°–4700°F, very little erosion occurs. Once this temperature is exceeded, the erosion rate increases greatly and can double, or even triple, for each 100°F temperature rise above this. The thermal properties, at high temperatures, are well enough known to make the design of a rocket nozzle possible. Figure 36 shows schematically the design and heat-flow concept of such a nozzle, and in Fig. 37 (A and B) a typical washer nozzle is shown before and after firing. The major engineering problem with nozzles of this type has been in coping with the unusually large expansion that occurs in the axial direction. Many methods have been developed for handling this problem and all seem to work quite satisfactorily. In general, present designs limit the length of the pyrolytic graphite throat insert to areas where very severe erosion occurs, and utilize conventional graphite or plastics in entrance and exit areas where erosion is less severe. In tests made with nozzles fired under identical conditions, pyrolytic graphite has shown obvious advantages in erosion resistance over conventional graphite and the high-density impregnated or hot-pressed varieties.

Another design utilizing the basic heat-sink concept is the wedge design sketched in Fig. 38. A number of these have been built and tested with remarkable success, using both very high temperature solid and liquid fuels. This design eliminates the large axial expansion problem associated with the washer design and substitutes for it large expansions

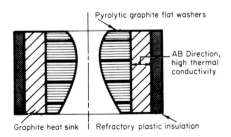

Pyrolytic graphite flat washers

AB Direction, high thermal conductivity

Graphite heat sink | Refractory plastic insulation

FIG. 36. Pyrolytic graphite stacked washer nozzle design. The stacked washer design acts as a heat pump and is generally backed up with a graphite heat sink and insulation.

in the circumferential direction. Some very ingenious designs have been evolved by the Curtiss-Wright Corporation and others to solve this problem.

As a result of a large program undertaken by producers as well as users of pyrolytic graphite, the Polaris second-stage nozzle now utilizes pyrolytic graphite as a throat material. The Allegheny Ballistics Laboratory of Hercules Powder Company was a prime factor in this development.

Work has also continued on the development of relatively small thin-walled rocket thrust chambers of pyrolytic graphite. Rather than try to develop units for high-thrust, high-pressure application, investigators have turned their attention to light-weight, radiation-cooled, attitude-control chambers, in which the entire nozzle, including the combustion chamber, throat, and exit cone, is a single thin shell of pyrolytic graphite.

Improved production techniques have increased the probability of obtaining defect-free chambers, and progress has been made in the areas of high-strength "alloys" of pyrolytic graphite and post-deposition heat treatments. Improved specifications, adequate proof testing and ingenious injector design has resulted in the production of satisfactory 100-pound thrust chambers. Work has continued to develop 500-pound and 1000-pound thrust chambers.

In the future, pyrolytic graphite throats using the washer concept are expected to find wider and wider utilization. As the price of pyrolytic graphite has decreased, many of the price advantages associated with the refractory metals and hot pressed graphites have been overcome. For very small nozzles, where machining casts are comparable with raw material costs, pyrolytic graphite is even challenging conventional graphites. For very large throats, i.e., above 12 inches in diameter, cost and manufacturing difficulties will probably preclude the use of pyrolytic graphite for some time (61,62).

C. Nuclear Applications

The nuclear applications of pyrolytic graphite have been in three main areas. The first was the infiltration of graphite with pyrolytic "carbon" to render it gas tight. Work on infiltration was done in the 1950's in England by the United Kingdom Atomic Energy Authority (6–8). In the United States, work was done by several commercial companies interested in gas-cooled reactors and by the Atomic Energy Commission. Although considerable success was obtained in producing impervious graphites by infiltration, other more economical methods were soon developed. Apparently no major market exists for the products at present, although a limited market may exist for fuel-cell applications. Infiltration is generally carried out at lower pressures and temperatures than those

FIG. 37A. Typical stacked washer nozzle: before firing.

used for pyrolytic graphite deposition. These lower temperatures of deposition yield pyrolytic carbon rather than graphite.

The second major area of application in the nuclear field has been for coating power reactor fuel particles. The carbide or oxide particles are coated in a fluidized bed. Such pyrolytic "carbon-coated" fuel particles are incorporated in the high-temperature gas-cooled reactor (HTGR) and the pebble bed reactor (PBRE) in this country and the Dragon reactor in England. A wide range of deposit microstructures can be produced by controlling temperature, pressure, and residence time of the particles in the fluidized bed. Certain structures are better for fission-gas retention, whereas others show better thermal properties and wear resistance. The process is economical, and as gas-cooled reactors

FIG. 37B. Typical stacked washer nozzle: after firing.

become increasingly attractive for power generation, we can expect very significant increases in this market for pyrolytic graphite coatings. A typical high-temperature fluidized-bed particle-coating apparatus for laboratory experiments is shown in Fig. 39.

The pyrolytic coatings over fuel particles perform a number of important functions. They are impervious, thereby preventing loss of fission gas. Being impervious, they also prevent destructive hydrolysis of some of the radioactive carbides used as fuel (as ThC_2). Being carbon, they are compatible with reactor graphite moderator materials. Finally, the coatings have a very low thermal neutron-absorption cross section.

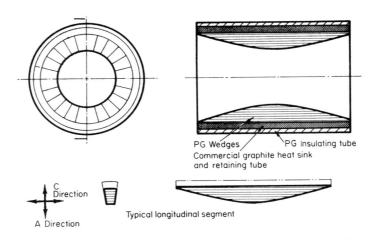

PG Wedges   PG Insulating tube
Commercial graphite heat sink
and retaining tube

C
Direction

A Direction

Typical longitudinal segment

FIG. 38. Wedge-design nozzle utilizing pyrolytic graphite (PG) plate.

Fission product retention for the lifetime of the fuel is now a popular research area. Studies are being made by Battelle Memorial Institute, General Atomics, and the Oak Ridge National Laboratory in the United States; the O.E.C.D. High-Temperature Reactor Project (Dragon), Nukem Nuklear-Chemii und Metallurgie GMBH, and Brown Boveri-Krupp in Europe; and Nippon Carbon Company in Japan. An extensive review has been published on nuclear applications of pyrolytic graphite (25).

The third major area of application is in nuclear rocket engines. The Rover project, a joint effort of the Atomic Energy Commission and the National Aeronautics and Space Administration, is designed to develop a nuclear propulsion system by the mid-1970's. Much of the original work was carried out by the Los Alamos Scientific Laboratories with their older KIWI series of reactors and newer Phoebus I and II high-powered series. In the early 1960's the effort was expanded to the NERVA (nuclear engine for rocket vehicle application) series by Aerojet-General Corporation and the Westinghouse Astronuclear Laboratory. A typical NERVA reactor is shown in Fig. 40. Although detailed discussions of the design are classified, it can be stated that the pyrolytic graphite in these reactors is used for high-temperature insulating purposes and not as a structure material. Some tens of thousands of parts have been fabricated from pyrolytic graphite for these reactors, from small tubes to large plates, and many are now considered "off-the-shelf" items. Much of the physical and thermal property data available on pyrolytic graphite was developed as a result of the Rover project. Recent cutbacks in the United States space program have significantly reduced the work being done in this field, but the unprecedented success that these reactors have had

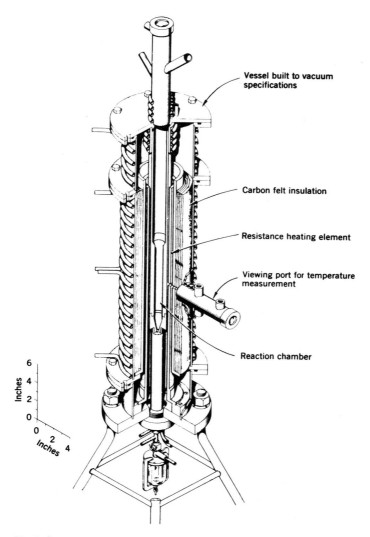

Vessel built to vacuum
specifications

Carbon felt insulation

Resistance heating element

Viewing port for temperature
measurement

Reaction chamber

Inches

6
4
2
0

0   2   4
Inches

FIG. 39. Laboratory apparatus for coating particles with pyrolytic graphite in a
fluidized bed at high temperatures (25).

in static tests should assure a good future market for pyrolytic graphite
in this area.

D. COMMERCIAL AND CONSUMER APPLICATIONS

The high cost of pyrolytic graphite has limited its use to aerospace
and nuclear markets. Pyrolytic graphite generally finds greatest applicability under very severe environmental conditions, which are seldom, if

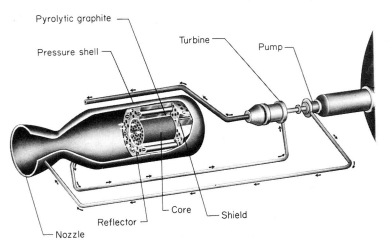

Fig. 40. Typical nuclear propulsion rocket motor.

ever, encountered in ordinary commercial and consumer areas. Ordinary commercial graphites at a hundredth to a thousandth of the cost are difficult to replace with pyrolytic graphite. The latter has found application in small parts, where machining and maintenance costs are much greater than the initial cost of the material.

Markets that have been examined include areas where the lower oxidation rates would be helpful, where anisotropy of thermal or electrical conductivity could make a significant contribution, or where the impermeability of pyrolytic graphite would be a great advantage. Electrical applications have been examined, including motor brushes, rheostat slide contacts, carbon resistor piles, etc. No significant market exists for these applications although the material is still being studied for new uses. Some small quantity of pyrolytic graphite is finding use in very high temperature furnaces as thermal insulation, and to give longer and more uniform heating zones due to the extremely high thermal conductivity in the AB plane. These furnaces are, of course, limited to reducing or vacuum conditions.

Attempts have been made to use pyrolytic graphite crucibles to melt and hold a number of very reactive materials. The impermeability of the material suggests that this could be a very significant market. However, high cost and other limitations have restricted this application.

Grafoil,[1] a pure, flexible insulating graphite tape with highly oriented properties similar to pyrolytic graphite and usually produced from an-

[1] Trademark of High Temperature Materials, Division of Union Carbide Corporation.

nealed pyrolytic graphite, has found some commercial and aerospace markets. The ability of this material to conform to shapes has great advantages; however, cost has limited the market.

The only known consumer product containing pyrolytic graphite has been a smoking pipe in which the bowl is lined with pyrolytic graphite (63). Several hundred thousand of these pipes have been sold and have been well received by the consumers. With this single exception, the general public is unaware that pyrolytic graphite exists.

VIII. Other Forms of Pyrolytic Carbon and Graphite

A. ALLOYS

When pyrolytic carbon or graphite is codeposited with other elements (e.g., boron, tungsten, silicon, hafnium, tantalum, molybdenum, niobium, zirconium, and titanium) the resultant mixture is called a pyrolytic alloy. The major emphasis in the fabrication of these materials has been in the development of alloys in elemental concentration of up to two atom percent. Boron pyrolytic graphite is virtually the only one of the material produced in the unclassified alloy programs that is available for engineering applications.

In general, the addition of small amounts of metals to pyrolytic graphite significantly alters the properties of the material. Considerably more effort is required with the alloy codeposition to achieve consistent high-quality results. High concentrations of additives alter properties more radically than small concentrations, but unfortunately, few data on these are available.

Pyrolytic graphite alloys fall into two broad categories, (a) low-concentration, single-phase, solid-state solution alloys (solution alloys) and (b) high-concentration multiphase alloys. The boron pyrolytic graphite alloy is a solution alloy.

To produce alloys, a gaseous or vapor-phase alloy metal halide is metered into the common gas injector with the hydrocarbon source gas and introduced into the furnace at 1900°–2300°C at 1–100 torr pressure. In the case of boron pyrolytic graphite, the ratio of boron to carbon in the gas can be varied over a small range about the desired ratio of boron to pyrolytic graphite, mainly as a function of deposition temperature and pressure. A good summary of boron pyrolytic graphite alloy properties may be found in a paper by S. F. D'Urso, presented at the ASTM Symposium on Pyrolytic Graphite in March 1964 at Palm Springs, California.

The presence of 0.3% to 1.0% boron has the following effects on pyrolytic graphite:

1. No substantial change in C-direction thermal conductivity.

2. A reduction of AB-direction thermal conductivity in a temperature range up to 1400°F (760°C).

3. An increase in ultimate tensile strength up to 4000°F (2205°C).

4. An increase in fracture elongation and overall ductility.

5. An increase in the degree of graphitization as deposited.

6. An impedance in the transformation due to heating only.

7. A process advantage in the manufacture of certain types of hardware (notably, very high wall-thickness-to-radius ratios, t/r 0.18).

The effect of the alloy on bend strength and electrical resistivity properties suggests the probability that selected properties can be altered with certain other alloy systems. Some data have shown that molybdenum in a solution alloy approximately doubles the C-direction tensile strength. For multiphase alloys, the process becomes more complex, since greater concentrations of the alloy gas must be used. A detailed discussion of these problems has been given (64). Since the multiphase alloys are mixtures of metal carbide and pyrolytic graphite, the gas concentrations must be held with carbon in excess of the stoichiometric ratio of metal to carbon in the carbide. Typical heterogeneous equilibrium-phase problems complicate the process, such as low-melting eutectics, new carbide compounds that form solid solutions with carbon, and hypereutectic carbon precipitation phases. Work at Raytheon on ZrC alloys (64) developed a marked increase over unalloyed pyrolytic graphite in strength (threefold at room temperature), which held to 4000°F. In the High Temperature Materials program a B_4C alloy of pyrolytic graphite was prepared which showed not only an increase in strength but a doubling of the room-temperature modulus of elasticity (to 8.2×10^6 psi).

B. Foam

It was discovered many years ago (66) that small crystals of graphite could be treated in such a way that they could be exfoliated in the C direction with substantially no dimensional change in the AB direction. The treatment consisted of soaking the graphite in a solution of aluminum chloride and sulfuric acid for short periods, followed by drying gently and heating to about 500°C (932°F). On heating, the material expanded in a manner resembling the familiar "4th of July snake," the expansion in the C direction being of the order of several hundred to a thousand percent. The process was limited to very small pieces since large crystals of graphite of over one-half inch size are virtually never found in nature. Shortly after it became possible to produce massive shapes of pyrolytic

FIG. 41. Foamed graphite.

graphite, it was discovered that these could be converted to large single crystals of graphite by heating to 3000°C (5432°F). When these large crystals were subjected to the same process, large sections of exfoliated graphite several inches in diameter could be produced.

Foamed graphite is sold under a variety of trade names. It has not, however, found wide applicability in the market place and is still somewhat of a laboratory curiosity. Several pieces of this material are shown in Fig. 41. The foam so produced has even greater thermal and electrical anisotropy than conventional pyrolytic graphite. However, it is extremely weak structurally, and in the completely expanded form will not support loads of even a few pounds per square inch. Pyrolytic graphite foam can be compressed into a variety of shapes which will support greater loads, but at a corresponding loss in anisotropy. At very high compressions it resembles a large crystal of graphite and is weaker and has less anisotropy than pyrolytic graphite. To use the foam at high temperatures, it is necessary to heat the material to above 1650°C (3000°F) in vacuum to remove all of the reaction products.

C. FOIL

A brief description of foil graphite was given in the section on consumer applications for pyrolytic graphite. The starting point for the production of foil is foamed pyrolytic graphite or purified foamed natural graphite flake. In either case the foamed material with a very small amount of binder is passed through a series of rolls until a well-consolidated strip is produced. As a result of its natural tendency to shear along the basal (AB) plane, the material has a preferred crystallographic orientation similar to pyrolytic graphites or single-crystal graphite. Tapes, several inches wide and up to 100 inches long, are available in a variety of thicknesses.

D. ANNEALED PYROLYTIC GRAPHITE

Massive pieces of pyrolytic graphite have been converted to large, nearly perfect single crystals by heating to temperatures in excess of 2900°C (5252°F) and holding for several hours. Even more perfect crystals can be made if the material is subjected to a strain while the heating is being done. During this process the bulk material shrinks in the C direction about 11%, as the lattice parameter C_0 decreases 1.9%. In the AB direction the material grows 3–4%, and the corresponding volume change raises the density from 2.20 to 2.26 gm/cc. Pyrolytic graphite so treated is extremely soft and can easily be scratched with the fingernail. It is possible to peel thin layers of material from the surface due to its very low C-directional tensile and shear strength. These large crystals of high purity graphite have found some use in aerospace applications. They have also been used for single-crystal standards in

Fig. 42. Typical structures made from felt-based reinforced pyrolytic graphite.

studying radiation damage in graphites and have been used as X-ray monochromator crystals.

E. REINFORCED PYROLYTIC GRAPHITES

With a wide variety of fibrous carbonaceous substrates it has been found possible to convert to carbon, infiltrate and build up massive bodies of reinforced pyrolytic carbon and graphite that are essentially isotropic in properties. Such a family of materials, known as RPG[2] carbon materials have been produced using cross-needled precursor fibers carbonized to felted carbon fibers and used as deposition substrates. The deposition process is carried out at a relatively low temperature, 980°C–1370°C (1800°F–2500°F). An entire range of materials is formed by stopping the gaseous pyrolytic coating of the fibers at any stage from the "as carbonized" precursor at 6-8 lb/ft³ or 0.1 gm/cc to virtually full

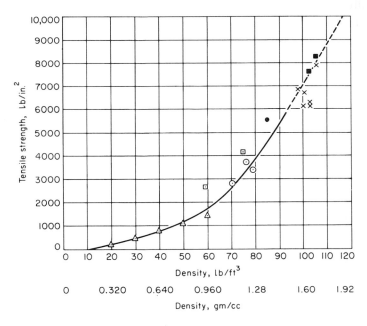

FIG. 43. Tensile strength of reinforced pyrolytic graphite at room temperature, AB direction. KEY: △, Super temperature data; ▢, Sandia test data; ☉, derived from Martin Co. flexure test data; ●, GE data-RT; ✕, GE data 2000°F; ■, GE data 3000°F.

2 Registered trademark of Super-Temp Company, Division of Ducommun, Inc., Santa Fe Springs, California, for reinforced pyrolytic graphite.

density at 120 lb/ft³ or 1.92 gm/cc. The structure of the product formed is that of pyrolytic carbon. However, the anisotropy is localized, relative to the deposition surfaces, and the overall body is more nearly isotropic. The theoretical density of this carbon is 2.05 gm/cc, so that 1.92 gm/cc represents a 95% dense structure.

These structures can be graphitized by heating to 2750°C or above. The density-dependent properties of RPG carbon materials depend greatly on the initial bulk density of the fibers, the denier, density, and fiber length of this substrate, the temperature of the deposition process, the degree of open porosity, the amount of pyrolytic carbon the final product contains, and subsequent heat treatments. A particularly interesting feature of RPG materials is that at any given density the available bulk porosity is virtually 100% open. This is considered to be caused by the porosity following the fibers into the structure. The open character of the porosity is convenient for further infiltration or impregnation with alloy elements

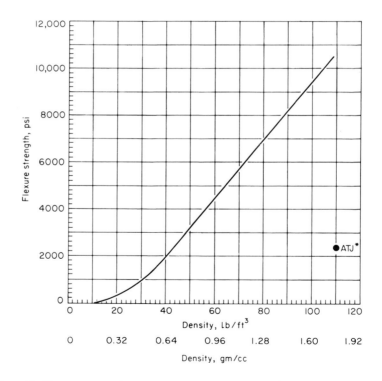

FIG. 44. Flexure strength at room temperature of reinforced pyrolytic graphite, plotted against density in the AB direction. The formula is flexure strength (psi) = 126 × density (lb/ft³) — 3070. This applies from 40 to 110 lb/ft³.

or densification with pore-blocking resins, or both.

Substrates that have been infiltrated with pyrolytic "carbon," in addition to felts, include "3-D Weaves," and charred carbon phenolic mixtures, such as RPP,[3] Carbitex[4] (a filament-wound material), and Pyrocarb.[5] With the exception of the felts, the resulting structure consists of 75% or more of the initial substrate, and the pyrolytic "carbon" is the glue which holds the structure together. In the felts, the initial substrate can be from 75% to as low as 4% of the final weight. Some typical structures made from these felts are shown in Fig. 42.

Uniform parts up to several feet in cross section have been produced from felt-based RPG, and in thicknesses up to 6 inches, with a high measure of three-dimensional strength (anisotropy of only 2 to 1 in com-

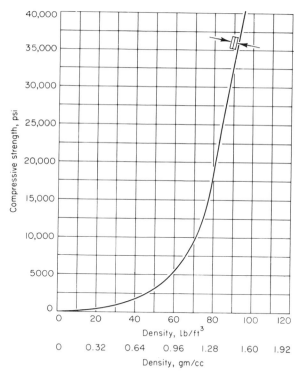

Fig. 45. Compressive strength at room temperature of reinforced pyrolytic graphite, plotted against density in the C direction.

[3] Trademark of Ling Temco Vought Aerospace Corporation, Dallas, Texas.
[4] Trademark of The Carborundum Company, Niagara Falls, New York.
[5] Trademark of HITCO Corp., Gardena, California.

pression). With other reinforced pyrolytic graphites, thickness is generally limited to under 1 inch owing to the difficulty of penetrating the tighter structures without sealing off the outer layers. Typical tensile, flexure, and compressive curves of a felt-base material as a function of density are shown in Figs. 43, 44, and 45.

Using very large furnaces these materials can now be made less expensively than pyrolytic graphite since the process is carried out at relatively low temperatures it appears possible to reduce the price further if a large demand should develop. Because all the forms of reinforced pyrolytic graphite begin with open porosity, alloying elements such as niobium, tantalum, zirconium, silicon, boron, and hafnium may be incorporated in concentrations up to 33% by weight of the RPG carbon materials. The wide variety of reinforced pyrolytic graphite products makes them impossible to describe fully. The reader is referred to the companies mentioned for a further description of the products now available.

Acknowledgments

The writers would like to acknowledge the work of J. D. McClelland of Aerospace Corporation who spent countless hours abstracting and reviewing literature and assisting in writing of the text. They would further like to acknowledge the contribution of the many workers from whose literature they borrowed. Final acknowledgment is due Rosemary Hayduk of Super-Temp Company for assembling, editing, and putting this chapter together, and to Karen Robért for her assistance in the typing of the chapter.

References

1. C. F. Powell & Associates, "Vapor Deposition" Wiley, New York, 1966.
2. W. L. Voelker, Manufacture of filaments for incandescing electric lamps, U. S. Patent No. 683,085 (September 24, 1901).
3. W. E. Sawyer and A. Man, U. S. Patent No. 229,335 (June 27, 1880).
4. Erle I. Shobert II, Carbon and graphite. *Modern Materials* 4, 1–99 (1964).
5. R. O. Grisdale, A. C. Pfister, and W. vanRoosbroeck, Pyrolytic film resistors: Carbon and borocarbon. *Bell System Tech. J.* 30, 271–314 (1951).
6. A. R. G. Brown and W. Watt, The preparation and properties of high temperature pyrolytic carbon. *In* "Industrial Carbon and Graphite," pp. 86–100. Soc. Chem. Ind., London, 1958.
7. A. R. G. Brown, A. R. Hall, and W. Watt, Density of deposited carbon. *Nature* 172, 1145 (1953).
8. A. R. G. Brown, D. Clark, and J. Eastabrook, Some interesting properties of pyrolytic graphite. *J. Less-Common Metals* 1, 94–100 (1959).
9. R. Gomer and L. Meyer, Energy exchange between cold gas molecules and a hot graphite surface. *J. Chem. Phys.* 23, 1370 (1955).
10. R. J. Diefendorf, The deposition of pyrolytic graphite. *J. Chem. Phys.* 30, 815 (1960).
11. E. Voice, The pyrolysis of methane in a bed of particles. *O.E.C.D. High Tem-*

perature Reactor Project Dragon, D. P. Rept. No. 569, Atomic Energy Establishment, Dorset, England (May, 1967).

12. P. H. Higgs *et al.,* Studies of graphite deposited by pyrolytic processes. *Wright Air Development Div. Tech. Rept. No. TR 61–72,* pp. 37, 82 and 85 (May, 1964).

13. R. E. Duff and S. H. Bauer, The equilibrium composition of the C/H system at elevated temperature. *U. S. Atomic Energy Comm., Rept. No. LA-2556* (June, 1961).

14. J. H. Richardson and E. H. Zehms, Structural changes in pyrolytic graphite at elevated temperatures. Aerospace Corporation, El Segundo, California, U.S.A.F. Space Systems Division, Technical Documentary Rept. No. 63-340, Contract No. AF 04(695)-269 (December, 1962).

15. E. R. Stover, Microstructure of pyrolytic graphite. Paper presented at ASTM Symposium on pyrolytic graphite, Palm Springs, California (March, 1964).

16. E. R. Stover, Effects of annealing on the structure of pyrolytic graphite. *General Electric Res. Lab., Rept. No. 60-RL-2564M* (November, 1960).

17. E. R. Stover, Mechanisms of deformation fraction in pyrolytic graphite. *General Electric Res. Lab., Rept. No. 61-RL-2745M* (June, 1961).

18. E. R. Stover, Structure of whiskers in pyrolytic graphite. *General Electric Res. Lab., Rept. No. 64-RL-3609M* (March, 1964).

19. W. V. Kotlensky and H. E. Martens, Mechanical properties of pyrolytic graphite to 2800 C. *In* "Proceedings of the Fifth Carbon Conference" (S. Mrozowski, M. L. Studebaker, and P. L. Walker, Jr., eds.), Vol. 2, pp. 625–638. Pergamon, Oxford, 1963.

20. J. Pappis and S. L. Blum, Properties of pyrolytic graphite. *J. Am. Ceram. Soc.* **44,** 592–597 (1969).

21. O. J. Guentert and C. T. Prewitt, X-Ray diffraction study of pyrolytic graphite. Paper presented at the American Physics Society Conference, Detroit, Michigan (March, 1960).

22. R. Iley and H. L. Riley, The deposition of carbon on vitreous silica. *J. Chem. Soc.* **11,** 1362–1366 (1948).

23. L. F. Coffin, Jr., Structure-property relations for pyrolytic graphite, *J. Am. Ceram. Soc.,* **47,** 473–478 (1964).

24. Pyrolytic graphite—property data. High Temperature Materials, Inc., Lowell, Massachusetts, 1962.

25. J. E. Hove and W. C. Riley, "Modern Ceramics—Some Principles and Concepts." Wiley, New York, 1965.

26. W. V. Kotlensky and H. E. Martens, Tensile properties of pyrolytic graphite to 5000 F. *Jet Propulsion Lab., California Inst. Technol., JPL Rept. No. 32-71* (1961).

27. R. N. Donadio and J. Pappis, The mechanical properties of pyrolytic graphite. *Raytheon Company, Waltham, Massachusetts, Rept. No. T-574* (undated).

28. H. E. Davis *et al.,* "The Testing and Inspection of Engineering Materials." McGraw-Hill, New York, 1955.

29. J. Pappis, Mechanical properties of pyrographite. *In* "Mechanical Properties of Engineering Ceramics" (W. W. Kriegel and H. Palmour, eds.), Wiley (Interscience), New York, 1961.

30. J. Gebhardt and J. M. Berry, Mechanical properties of pyrolytic graphite. *AIAA J.* **3,** 302–308 (1965).

31. J. E. Dorn, The spectrum of activation energies for creep, *in* "Creep and Recovery." Am. Soc. Metals, Metals Park, Ohio, 1957.

32. H. E. Martens, L. D. Jaffe, and D. D. Button, High temperature short-time creep of graphite. *Trans. Metallurgical Soc. AIME* **218**, 782–787 (1960).

33. H. E. Martens *et al.,* Tensile and creep behavior of graphites above 3000 F. *In* "Proceedings of the Fourth Conference on Carbon" (S. Mrozowski, M. L. Studebaker, and P. L. Walker, Jr., eds.), pp. 511–530. Pergamon, Oxford, 1959.

34. J. J. Cacciotti and V. J. Erdeman, Fatigue behavior of a pyrolytic graphite. *General Electric Company, Aircraft Gas Turbine Division, Rept. No. DM61-229* (July, 1961).

35. "Pyrolytic Graphite Data Sheets." High Temperature Materials, Inc., Lowell, Massachusetts (February, 1962).

36. J. E. Hove and W. C. Riley, "Ceramics for Advanced Technologies." Wiley, New York, 1956.

37. N. S. Rasor and J. C. McClelland, Thermal properties of materials, Part 1, Properties of graphite, molybdenum and tantalum to their destruction temperatures. *Wright Air Development Division Tech. Rept. No. 56–400* (October, 1956).

38. C. A. Klein, Pyrolytic graphite. *Intern. Sci. Tech.* pp. 60–68 (August, 1962).

39. G. Slack, Anisotropic thermal conductivity of pyrolytic graphite, *Phys. Rev.* **127**, 3 (August, 1962).

40. W. Bradshaw and J. R. Armstrong, Pyrolytic graphite, its high temperature properties. Lockheed Missile and Space Company, Sunnyvale, California, *USAF Applied Systems Division-Tech. Documentary Rept. No. 63-195,* Wright-Patterson Air Force Base, Ohio (March, 1963).

41. W. W. Lozier, Development of graphite and graphite base multi-component materials for high temperature service. *Wright Air Development Division Tech. Rept. No. 789* (April, 1960).

42. W. De Sorbo, Low temperature values from 28 K to 300 K for Ceylon natural graphite, *J. Chem. Phys.* **26**, 244–247 (1957).

43. G. B. Spence, Electron microscope studies of vacancy controlled processes in graphite. *Wright Air Development Division Tech. Rept. No. 61-72,* 2 (1963).

44. J. D. Plunkett and W. D. Kingery, The spectral and integrated emissivity of carbon and graphite. *In* "Proceedings of the Fourth Conference on Carbon," pp. 457–481. Pergamon, Oxford, 1960.

45. R. J. Champetier, Basal plane emittance of pyrolytic graphite at elevated temperatures. Space and Missile Systems Organization, Air Force Systems Command, *Los Angeles Air Force Station, Tech. Rept. No. 0158* (3250-20)-10 (July, 1967).

46. R. J. Thorn and O. C. Simpson, Measurements of spectral emissivity at 6500 Å. *J. Appl. Phys.* **24**, 633 (1953).

47. C. A. Klein, Pyrolytic graphites: Their description as semimetallic molecular solids, *J. Appl. Phys.* **33**, 3338 (1962).

48. A. R. Ubbelohde and F. A. Lewis, "Graphite and Its Crystal Compounds." Oxford Univ. Press, London and New York, 1960.

49. C. A. Klein, Electrical properties of pyrolytic graphite. *Rev. Mod. Phys.* **34**, 56 (1962).

50. G. Wagoner and B. H. Eckstein, Research and development on graphite. *Wright Air Development Division Tech. Rept. No. 61-72,* 20 (November, 1963).

51. C. A. Klein *et al.,* Low temperature studies of magneto-resistive effects in the

basal planes of pyrolytic graphite annealed at 3600 C. *Phys. Rev.* **125**, 468–470 (1962).

52. F. P. Bundy and E. R. Stover, Electrical resistance of pyrolytic graphite under high pressures. *General Electric Res. Lab. Memo Rept. No. MF-1095* (January, 1962).

53. J. W. McClure and Y. Yafet, Theory of the g-factor of the current carriers in graphite single crystals. *In* "Proceedings of the Fifth Carbon Conference." (S. Mrozowski, M. L. Studebaker, and P. L. Walker, Jr., eds.), Vol. 2, pp. 22–28. Pergamon, Oxford, 1963.

54. W. S. Horton, Oxidation kinetics of pyrolytic graphite. *In* "Proceedings of the Fifth Carbon Conference." (S. Mrozowski, M. L. Studebaker, and P. L. Walker, Jr., eds.), Vol. 2, pp. 233–241. Pergamon, Oxford, 1963.

55. S. Levy, Thermal stresses in pyrolytic graphite. AIAA Structures and Materials Meeting, March, 1962.

56. J. G. Campbell *et al.*, Free standing pyrolytic graphite thrust chambers for space operation and attitude control. *The Marquardt Corporation, Van Nuys, California, Project 5016, Rept. No. 6106*, AFRPL-TR-66-95, under contract AF 04(611)-10790 (June, 1966).

57. A. M. Garber and E. J. Nolan, Thermal stresses in pyrolytic graphite during sustained hypersonic flight. *In* "Dynamics of Manned Lifting Planetary Entry," p. 536. Wiley, New York, 1963.

58. T. J. Clark and R. J. Larsen, Residual stress studies. *General Electric Company (Detroit) Rept. No. P1-WS-0232*, MPD (1961).

59. W. H. Smith, Effect of structures on strength at room temperature. Pyrolytic graphite information folder. *General Electric Company, Memo Rept. No. MF-109(b)F.*

60. S. M. Scala and L. M. Gilbert, Aerothermochemical behavior of graphite at elevated temperatures. *General Electric Company (Philadelphia) Rept. No. R635D89; AD 432*, 594 (November, 1963).

61. J. J. Cacciotti, Comparison of the physical properties and rocket firing performance of several graphite materials: ATJ, pyrolytic and ZT grades. *General Electric Company, Aircraft Gas Turbine Division, Rept. No. DM60-232* (October, 1960).

62. R. Francis, Development of pyrolytic refractory materials for solid fuel rocket motor applications. Arthur D. Little, Inc., Cambridge, Massachusetts, *Final Rept. to Watertown Arsenal under Contract DA-19-020-ORD-5490, Boston Ordnance District* (July, 1962).

63. U. S. Patent No. 3,185,163, May 25, 1965 and U. S. Patent No. 3,420,244, January 7, 1969.

64. Raytheon Corporation, Task 111, Pyrocarbides research and development. *Final report under contract NOrd 19135 (FBM)* (January, 1963).

65. R. J. Diefendorf, Preparation and properties of pyrolytic graphite. *General Electric Res. Lab., Rept. No. 60-RL-2572M* (November, 1960).

66. R. C. Croft, New molecular compounds of graphite, *Research* **10**, 23 (1957).

67. C. L. Mantell, "Handbook of Carbon and Graphite," Wiley (Interscience), New York, 1968. (A recent general reference.)

MATERIALS FOR TEMPERATURE MEASUREMENT

J. C. Chaston

Consultant; Formerly Manager, Research Laboratories, Johnson Matthey & Co., Ltd.,
London, England

I. Introduction

Heat cannot of itself pass from a colder to a hotter body
(R. J. E. Clausius, 1850, Second Law of Thermodynamics)

A. THE TEMPERATURE SCALE

Materials play a greater part in the measurement of temperature than in the measurement of any other basic property. The reasons for this are inherent in the nature of temperature itself, and warrant some preliminary discussion.

Temperature differs from other basic properties in that it cannot be measured simply by adding unit degree to unit degree. An unknown mass can be balanced against a sufficient number of unit masses and its weight expressed as their sum. An unknown length can be marked out in multiples of one unit length. A period of time, even, can be measured by counting the unit pulses—the heart beats, the swings of a pendulum, or the waves of light received from a standard source—which tell out its span.

But if we put together two identical bodies at the same temperature, as by mixing two equal amounts of liquid, then, although the mass will be doubled, the volume doubled, and the heat content doubled, the temperature will remain the same. More formally, it may be said that temperature is an *intensive* quantity, in contrast with other *extensive* quantities in our measuring system.

The one simple thing that can be done to evaluate temperatures of bodies at different unknown temperatures is to *order* them by the application of the second law of thermodynamics; that is, to determine, by the direction of heat flow, which is the hotter and which the colder of each pair of bodies.

When it comes to establishing a scale of temperatures, however, the only practical approach is to make use of the temperature-sensitive characteristics of materials. At first sight, it would appear that the selection must be arbitrary and that the scale we define must equally be arbitrary; for, to quote Wensel (1), "the size of a degree on one part of a scale, *no matter how defined,* can bear no relation to the size of a degree on any other part of the scale." He goes on to say, by way of example,

We can define a scale by making equal increments on the scale correspond to equal increases in the length of a tungsten rod. Thus 50° would be the temperature at which the length of the rod is halfway between its length at 0° (in melting ice) and its length at 100° (in steam). Are the degree intervals on this scale equal? Experience shows that the length of a rod of copper will be halfway between its length at 0°

and its length at 100° at a temperature which on our tungsten rod expansion centi-
grade scale is 54°. If we divide up the interval in proportion to the change in
electrical resistance, the *resistance* scale based on copper will agree with the
expansion scale based on copper to within a very small fraction of a degree. The
resistance scale based on tungsten agrees with none of the other three. Which
property divides the scale equally? Which material? Aside from this, the linear function chosen, even though the simplest choice, is entirely arbitrary.

The dilemma which thus arises can fortunately be resolved theoretically by reference to thermodynamic reasoning, and it is possible to
establish a theoretical scale, the Thermodynamic Temperature Scale of
Kelvin and Clausius, which is absolute in character. This is not the place
to discuss the mathematical derivation of the Kelvin scale, but essentially
the scale may be taken as being equivalent to one based on the expansion of an ideally perfect gas under constant pressure, the volume of the
gas expanding by equal amounts for each unit change in temperature
over the whole region of the scale from absolute zero upwards.

Fortunately, nitrogen is very nearly a perfect gas in its expansion
behavior, and it is possible to apply corrections for most of its deviations
from the ideal over a very wide range in temperature. Thus, by using
it in a gas thermometer, a means is available for constructing a practical
scale of temperature measurement.

It will, however, be appreciated that a gas thermometer is by no
means simple to use and can have few, if any, commercial applications.
It is bulky, fragile, and slow to come to equilibrium. The bulb of necessity needs to be of reasonable volume and thus can only be applied to
measure accurately a relatively large source of uniform temperature.
Such an instrument, carefully used in skilled hands, does however provide a means of establishing the temperatures of a series of reference
sources of constant temperature; and on this basis it has been used to
derive the universally accepted Practical International Scale of
Temperature.

The use of fixed reference points, which thus becomes the basis of
practical temperature measurement, seems first to have been suggested
by H. Fabri of Leyden in 1669. The Swedish astronomer, Anders Celsius,
whose name is commemorated in the units of the Practical International
Scale, is generally given the credit of proposing, in 1762, the now familiar
primary standards of melting ice and of steam at atmospheric pressure
(although curiously enough, to our way of thinking, he denoted the
steam point as 0°C and the ice point as 100°C!).

Both these reference points, which have formed the basis of our
practical temperature scale for over two centuries, represent phase transition temperatures in pure materials; and such transition points, in fact,

have been found to be the most convenient and dependable of fixed points for temperature measurement. Some are more convenient and reliable than others, however, and in recent years a good deal of attention has been given to studying the conditions affecting their reproducibility.

The International Practical Temperature Scale (I.P.T.S.) was introduced in 1927. A revision in 1948 altered the values of temperatures in the higher range, and a significant revision, with an extension of the whole range, was agreed on in 1968 and adopted by the standardizing laboratories of the world on January 1, 1969.

The scale is based on two fundamental points, the melting point of ice, 0°C or 273.15 K, and the boiling point of water, 100°C under standard atmospheric pressure; it also lays down a number of "defining fixed points." These derive from most careful thermometer determinations by the various standardizing bodies throughout the world—the Bureau of Standards in Washington, the National Physical Laboratory in London, the Bureau of Weights and Measures in Sèvres, and their equivalent organizations in Germany and Russia. The work is exacting and difficult and not always in as good accord as might be wished. Thus, until 1968 the principal results that determined the critically important "gold point," the freezing point of gold, were those of Holborn and Day, who in 1900 found a value of 1064°C, and of Day and Sosman, who in 1911 gave it as 1062.4°C. The figure chosen as a "defining fixed point" was 1063°C. However, in 1968 this was changed to 1064.43°C, largely as a consequence of determinations made in the intervening years at Brunswick in Germany and at Leningrad in Russia, as well as of determinations made before the war at the Massachusetts Institute of Technology.

The complete list of the present "defining fixed points" that govern the International Practical Temperature Scale 1968 is given in Table I in degrees Celsius (°C) and degrees Kelvin (now to be written simply without the superscript as K) and compared with those for the 1948 scale. All determinations are made at standard atmospheric pressure, unless otherwise specified. The triple point of water is adopted as a defining fixed point, as it is more accurately reproduced than the ice point, which is defined as 273.1500 K.

Besides defining the primary fixed points, the text of the International Practical Temperature Scale lays down methods by which the intervals between each shall be subdivided. These are

a. From 13.81 K (−259.34°C) to 630.74°C, a platinum resistance thermometer is to be used, using platinum sufficiently pure that the ratio of the resistance of the thermometer at 100°C to that at 0°C is at least

TABLE I

DEFINING FIXED POINTS OF INTERNATIONAL PRACTICAL TEMPERATURE
SCALES, 1968 AND 1948 REVISIONS

Fixed point	K (I.P.T.S. 1968)	°C (I.P.T.S. 1968)	°C (I.P.T.S. 1948)
Triple point of hydrogen	13.81		
Equilibrium point of hydrogen at 24/76 atmosphere	17.042		
Boiling point of hydrogen	20.82		
Boiling point of neon	27.102		
Triple point of oxygen	54.361		
(Triple point of argon)[a]			−189
Boiling point of oxygen	90.188		−182.970
Triple point of water	273.1600	0.0100	0.0100
Boiling point of water		100	100
(Boiling point of sulfur)[a]			449.600
Freezing point of zinc		419.58	
Freezing point of silver		961.93	960.8
Freezing point of gold		1064.43	1063

[a] These points have not been adopted for the 1968 scale.

1.3925:1. The thermometer is calibrated at the defining fixed points in this range and intermediate values are obtained by the use of two defined reference functions, one for temperature below 0°C and the other for the higher range.

b. From 630.74°C to the gold point the platinum:10% rhodium-platinum thermocouple is used, employing the relationship

$$E = a + bt + ct^2$$

where E is the emf in volts at temperature t°C, and a, b, and c are constants.

c. Above the gold point, a radiation thermometer is used, the results being calculated according to Planck's radiation formula, taking the value of the second radiation constant C_2 as 0.014388 meter Kelvin.

Since reference books and tables published before 1969 will continue to be used for many years, it is convenient to have a conversion table for the differences—generally quite small—between the two scales. This is given in Table II.

The details of the method used for establishing the International Practical Temperature Scale, although of some academic interest, are not of great practical importance, since they have been used once and for all to determine a series of secondary standards that are more than

TABLE II

Conversion Table Showing Differences Between Values for °C in the 1968 and 1948 Revisions of the International Practical Temperature Scale[a]

t_68 °C	0	-10	-20	-30	-40	-50	-60	-70	-80	-90	-100
-100	0.022	0.013	0.003	-0.006	-0.013	-0.013	-0.005	0.007	0.012	0.029	0.022
-0	0.000	0.006	0.012	0.018	0.024	0.029	0.032	0.034	0.033	0.029	0.022

t_68 °C	0	10	20	30	40	50	60	70	80	90	100
0	0.000	-0.004	-0.007	-0.009	-0.010	-0.010	-0.010	-0.008	-0.006	-0.003	0.000
100	0.000	0.004	0.007	0.012	0.016	0.020	0.025	0.029	0.034	0.038	0.043
200	0.043	0.047	0.051	0.054	0.058	0.061	0.064	0.067	0.069	0.071	0.073
300	0.073	0.074	0.075	0.076	0.077	0.077	0.077	0.077	0.077	0.076	0.076
400	0.076	0.075	0.075	0.075	0.074	0.074	0.074	0.075	0.076	0.077	0.079
500	0.079	0.082	0.085	0.089	0.094	0.100	0.108	0.116	0.125	0.137	0.150
600	0.150	0.165	0.182	0.200	0.22	0.25	0.27	0.30	0.32	0.35	0.38
700	0.38	0.40	0.43	0.46	0.48	0.51	0.54	0.57	0.60	0.62	0.65
800	0.65	0.68	0.71	0.74	0.77	0.80	0.83	0.86	0.89	0.92	0.94
900	0.94	0.97	1.00	1.03	1.06	1.09	1.12	1.15	1.18	1.21	1.24
1000	1.24	1.27	1.30	1.33	1.36	1.39	1.42	1.44	1.46	1.48	1.5

t_68 °C	0	100	200	300	400	500	600	700	800	900	1000
1000	1.5	1.5	1.7	1.8	2.0	2.2	2.4	2.6	2.8	3.0	3.2
2000	3.2	3.5	3.7	4.0	4.2	4.5	4.8	5.0	5.3	5.6	5.9
3000	5.9	6.2	6.5	6.9	7.2	7.5	7.9	8.2	8.6	9.0	9.3

[a] Example: To find the value of 250°C, I.P.T.S. 1968 in the I.P.T.S. 1948, subtract 0.061° from 250°, with the result 249.039.

TABLE III

SECONDARY STANDARDS OF INTERNATIONAL PRACTICAL TEMPERATURE
SCALE, 1968 AND 1948 REVISIONS

Standard	°C (I.P.T.S. 1968)	°C (I.P.T.S. 1948)
Temperature of equilibrium between solid CO_2 and vapor	−78.49	−78.52
Freezing point of mercury	−38.85	−38.87
Temperature of transition of solid sodium decohydrate	32.37	32.38
Triple point of benzoic acid	122.43	122.36
Freezing point of indium	156.624	156.611
Temperature of equilibrium between naphthalene and vapor	218.05	218.0
Freezing point of tin	231.967	231.913
Temperature of equilibrium between benzophenone and vapor	305.97	305.9
Freezing point of cadmium	321.107	321.032
Freezing point of lead	327.503	327.426
Temperature of equilibrium between mercury and vapor	356.66	356.58
Freezing point of		
antimony	630.7	630.5
aluminum	660.37	660.1
copper in reducing atmosphere	1084.5	1083
nickel	1455	1453
cobalt	1494	1492
palladium	1554	1552
platinum	1772	1769
rhodium	1963	1960
iridium	2447	2443
tungsten	3387	3380

sufficient for calibrating all temperature-measuring instruments used in science and industry.

The more important of these are given in Table III. In order to make full use of these standards, very great care is obviously necessary in observing the transition point. The melting points of Zn, Pb, Cd, Sn, and In were determined by McLaren (2) to an accuracy of better than 0.001°C; using a simple metal block furnace with the cooling rate adjusted so that the plateau temperature remained constant for 1 to 2 hours.

The effect of impurities in fixed point standards is often queried. McLaren found that a change in the freezing point of zinc was altered by 0.00002°C by less than 0.1 ppm of Cd, Fe, Pb, Sn or Te or by 1 ppm of Cu. In tin, a similar change in the freezing point is produced by less than 0.1 ppm of Ca, Cu, Fe, Ni, and Pb and by 1 ppm of Bi or Sb.

B. PRACTICAL METHODS OF MEASURING TEMPERATURE

In practice, it is not always convenient to use the methods laid down by the International Practical Temperature Scale for measuring temperatures in industry, nor is it often desirable to do so. The International Practical Temperature Scale defines in effect how a basic calibration is to be determined. Any number of stable secondary methods of temperature determination may be used in science or in industry, often with advantages of speed, convenience, and even accuracy in the particular conditions of measurement.

The material characteristics that may be employed in practical temperature measurement include thermoelectric effects between dissimilar materials; electrical resistivity; and thermal expansion. In addition, as mentioned above, phase-change points such as those associated with melting and boiling of metals, alloys, and compounds can provide temperature indications for fixed points on the temperature scale.

Most materials find their main use only over certain ranges of temperature, within which they exhibit their greatest sensitivity or stability. The upper temperatures at which materials can be used are obviously often limited by the onset of melting or reaction with their environment. The lower limit is more usually dictated by a loss of sensitivity.

It is thus convenient to consider temperature-sensitive materials in about six groups according to the ranges in temperature in which each finds its broadest field of use. The materials most commonly used, either alone as resistance elements or in combination as thermocouples or expansion devices are classified in this way in Table IV. In the sections which follow, as indicated in the table, some of the characteristics of these materials which particularly affect their performance in temperature measurement are discussed.

It should be emphasized that no pretense is made here of covering completely the extensive science and technology of temperature measurement. The intent is to consider only the *materials* involved in temperature measurement, with emphasis on the precautions that should be observed in their selection, preparation, and use.

The fields of radiation and optical pyrometry are not considered, since in these methods of temperature measurement the material characteristics involved are in most instances only those of the hot bodies under observation and are outside the control of the technician.

The subject of temperature measurement should never be regarded as static; it is under constant study and developments can be expected in all its branches. As an indication of the attention being given currently to this field, one of the main committees of the American Society for Testing and Materials, E-20, is on temperature measurement. Active sub-committees cover thermocouples, liquid-in-glass thermometers, radiation pyrometers, optical pyrometers, bimetallic thermometers, resistance thermometers, filled system thermometers, and miscellaneous thermometers.

II. Thermocouple Materials

Thermocouples are more widely used than any other materials for temperature measurement in industry, and in this section the composition and characteristics of most of the combinations that have been proposed and used are discussed.

Particular attention is given to platinum:platinum alloy thermocouples. Although they can provide a reliable and accurate means of measuring temperature over a wide range, this is true only if they are used with a full appreciation of the precautions needed to avoid the contamination that can so easily take place. A great deal of information is now available on the causes and effects of contamination of platinum thermocouples—more, indeed, than for other thermocouple materials— and it has been considered worthwhile to set this out in some detail.

A. PLATINUM:RHODIUM-PLATINUM

Thermocouples made by coupling wires of pure platinum with wires of either 10% rhodium-platinum or 13% rhodium-platinum alloys have a wide range of industrial application. They may be used for measuring temperatures in the ambient range from —50° to 100°C, and they are widely used in the medium high and high ranges from 100° to 1000°C, and from 1000° to 1600°C. The couples are not generally used for sub-zero temperature readings, since the emf developed is too low for accurate measurement and becomes zero at about —138°C.

An outstanding characteristic of platinum:rhodium-platinum thermocouples is the consistency that can be maintained in their calibration from batch to batch. Commercial practice is to guarantee the calibration of all new couples to be within 1°C of that given in the standard tables at temperatures up to 1100°C, to be within 2°C up to 1400°C, and to be within 3°C at higher temperatures. The couples are, moreover, unusually stable up to temperatures of at least 1200°C when heated in air so that they may be used without protection from oxidation; although it needs to be appreciated that at high temperatures they must be shielded from many forms of possible contamination.

TABLE IV

TEMPERATURE-SENSITIVE MATERIALS AND THEIR RANGE OF USEFULNESS

Temperature range	Thermocouples	Resistance elements	Expansion pairs
Very low temperatures (−273° to −253°C; 0 to 20°K)	Co-Au:Cu (II,O)[a] Co-Au:normal Ag (II,P) Chromel:alumel (II,E) Cu:constantan (II,F)	Ge (III,C) In (III,D) Pb (III,D) AuAg (III,D)	
Low temperatures (−253° to −50°C; 20° to 225°K)	Pt:Rh-Pt (II,A) Chromel:alumel (II,E)	Pt (III,A) Thermistors (III,D)	
Ambient temperatures (−50° to 100°C)	Pt:Ir-Pt (II,B) Pt:Rh-Pt (II,A) Chromel:alumel (II,E) Cu:constantan (II,F) Fe:constantan (II,G) Pd-Au:Ir-Pt (II,L) Platinel (II,M)	Pt (III,A) Ni (III,B)	Th-Hg in glass (IV,B) Hg in glass (IV,A) Organic liquid in glass (IV,D)
Medium high temperatures (100° to 1000°C)	Pt:Rh-Pt (II,A) Pt:Ir-Pt (II,B)	Pt (III,A) Ni (III,B)	Hg in glass (to 600°C) (IV,A) Ga in SiO_2 (IV,C)

Chromel:alumel (II,E)
Cu:constantan (to 800°C) (II,F)
Fe:constantan (II,G)
Pd-Au:Ir-Pt (to 800°C) (II,L)
Platinel (II,M)
Fibro Pt:Rh-Pt (II,D)

High temperatures
(1000° to 1600°C)

Pt:Rh-Pt (II,A)
Rh-Pt:Rh-Pt (II,C)
Chromel:alumel (to 1300°C) (II,E)
Fibro Pt:Rh-Pt (II,D)
Mo:W (II,H)
W-Re:W (II,I)

Pt (to ~1400°C) (III,A)

Very high temperatures
(above 1600°C)

Mo:W (II,H)
W-Re:W (II,I)
Ir:Rh-Ir (II,J)
SiC:graphite (II,K)

Al_2O_3 (III,D)

ᵃ The references in parentheses are to the sections of this survey.

1. Platinum

The platinum used for the pure platinum element of this thermocouple is usually specially refined for the purpose, and the laboratory control maintained over its production has been described by Betteridge and his co-workers (3). Platinum consolidated by powder metallurgy is not generally considered acceptable, and at one time a clause really directed against the use of such material was frequently inserted in specifications. This required that the wire should melt quietly to a bead when the end was inserted in the tip of an oxyhydrogen blowpipe, without any spitting or sparking. Powder metallurgy platinum was not believed to meet this requirement. However that may be, platinum wire for thermometry is now commonly drawn from bar derived from ingots melted in zircon or zirconia crucibles in an induction furnace. The spectrographic standard adopted is equivalent to a minimum purity of 99.998%

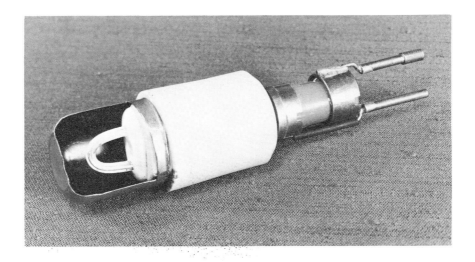

FIG. 1. Expendable fine wire thermocouple cartridge. The platinum:rhodium-platinum wires are seen in the U-shaped quartz tube within the sectioned steel protection cap. (Courtesy Leeds and Northrup.)

of platinum, but this in itself is not accepted as the criterion of suitability. The critical requirement is that the resistivity ratio R_{100}/R_0 shall be higher than 1.3920. The principal producers invariably work to a minimum of 1.3922.

Particular care is needed to avoid contamination of the metal while it is being worked to wire. Some surface pickup of iron is almost inevitable in forging, rolling, and wire-drawing operations, and it is especially important that this should be entirely removed by pickling before any annealing is attempted. Once the wire is annealed, any metallic impurity on the surface will quickly diffuse throughout the mass so that subsequent pickling is ineffective.

The standard wire diameter has for many years been 0.020 inch or 0.5 mm and in this diameter, when the wire is sheathed, the heat capacity of the tip and the thermal conductivity along the wire are both sufficiently low not to contribute any significant thermal lag to the assembly. The speed of response is virtually controlled by the heat capacity of the sheath. However, if bare couples can be used, there are obvious advantages in a thinner wire, and the development of the quick-reading single-immersion "throw-away" short couple (Fig. 1), heated so rapidly that contamination is arrested, has brought a demand for very thin elements, 0.002 inch in diameter, or even less. The production of these very thin wires requires extra cleanliness in drawing, with special attention to pickling and annealing programs.

2. Rhodium-Platinum Alloys

Two compositions are commonly used for rhodium-platinum alloys: 10% rhodium and 13% rhodium. These are nominal compositions, and the alloys are invariably made up so that the wires, when coupled with pure platinum, conform in their temperature-emf relationships with the nationally accepted standards. Experience shows that provided the alloys are made from thermocouple-grade platinum and pure rhodium, it is sufficient to calibrate a test wire from each batch at the gold point, 1064.43°C. If the composition is adjusted so that the emf is within ±1°C of the standard value, then the couple will conform equally closely with the standard at all other temperatures.

The persistence of two standard alloys is justified only by the continuing existence in industry of indicators and recording instruments calibrated for use with either 10% or 13% rhodium-platinum:platinum couples.

Technically, there is no justification for two alloys. The 10% rhodium-platinum alloy was first proposed by Le Chatelier in 1886, and was widely used until 1922, when as a result of an investigation of the stability of 10% rhodium-platinum:platinum couples on long heating, carried out at the Bureau of Standards, it was found that couples of European origin were contaminated by up to 0.34% of iron, derived from the rhodium used in their production. When, by improved refining methods, the iron was eliminated, the resulting reasonably pure 10% rhodium-platinum:platinum couples, while they retained their calibration much better in service at high temperature, developed an appreciably lower emf. The 13% rhodium-platinum:platinum couple was therefore introduced to match the old impure 10% couple, and to be interchangeable in existing pyrometer installations. Yet, such is tradition, so many specifications continue to demand the 10% alloy that the two alloys have remained in widespread use ever since.

3. Calibration

Two standard sets of calibration tables are in general use—one originating from U. S. National Bureau of Standards Circular 961, the other from British Standard 1826 (based on results from the National Physical Laboratory, England). Both refer to couples made nominally from either 10% or 13% of rhodium; but, as explained above, the alloys are actually made up to match the tables, and it is evident that the couples chosen (independently) as a basis for each of the two national standards actually differed by a small amount from each other in the average composition of their rhodium-platinum alloy elements. The full tables are so readily available from suppliers that they will not be reproduced, but Table V indicates the magnitude of the emf developed by the couples and also shows how little deviation there is between the two standard calibrations.

4. Stability in Service

In service, platinum:rhodium-platinum couples may change in calibration as a consequence of

a. Diffusion of rhodium across the junction

b. Loss of rhodium from the alloy wire by preferential formation of volatile rhodium oxide (when heated in free air)

c. Migration of rhodium from the rhodium-platinum element to the platinum element through the refractory

d. Contamination of the wires by impurity elements

a. Diffusion Across the Junction

Some diffusion across the junction must inevitably occur but very little reliable evidence is available as to its extent. In service, this almost

TABLE V

TEMPERATURE-EMF RELATIONSHIPS FOR PLATINUM: 13% AND 10%
RHODIUM-PLATINUM THERMOCOUPLES(I.P.T.S. 1948)
(U. S. and British Standard Calibrations)

| Temperature °C[a] | emf, mV | | | |
| | Pt: 13% Rh-Pt | | Pt: 10% Rh-Pt | |
	U. S. Circular 261	British Standard 1826	U. S. Circular 261	British Standard 1826
100	0.646	0.644	0.643	0.642
200	1.463	1.463	1.435	1.435
300	2.394	2.392	2.315	2.314
500	4.454	4.460	4.219	4.222
700	6.715	6.735	6.261	6.265
1000	10.471	10.510	9.570	9.591
1200	13.181	13.222	11.924	11.946
1300	14.562	14.617	13.120	13.155
1500	17.317	17.463	15.497	15.580
1600	18.679	18.855	16.673	16.770

[a] Cold junctions at 0°C.

certainly is not a significant factor in couple deterioration and, as will be evident from the following discussion, it is not easy to separate it from other forms of rhodium transport. In any event, diffusion across the junction in itself will not influence the calibration unless it is extensive. So long as the diffused zone is confined to the zone of uniform temperature surrounding the tip, it will act simply as a connecting link between the couple wires. Only if diffusion extends beyond the uniform zone will it influence the readings from the thermocouple by reducing the value of the emf developed.

b. Preferential Evaporation of Rhodium

When the platinum metals are heated in oxygen or in air to above about 800°C, they all lose weight more or less noticeably. The principal mechanism operating, except with palladium, is the formation of volatile oxides that diffuse through a boundary layer and then get carried away at a rate depending on the temperature and on the conditions in the surrounding atmosphere. (With palladium, the vapor pressure of the metal is, relatively, so high that evaporation of the metal itself accounts for the greater part of the observed loss in weight.)

In recent years, reasonably accurate data on the partial pressures of the gaseous oxides of platinum and rhodium have been determined by heating the two metals in controlled conditions in a moving stream of oxygen, and it would appear that at temperatures up to about 1400°C the values for RhO_2, the volatile rhodium oxide, and the corresponding platinum oxide, PtO_2, in one atmosphere of oxygen are of the same order. As the temperature is increased above 1400°C, however, the partial pressure of RhO_2 rises much more rapidly than that of PtO_2.

If these results are accepted, it would seem reasonable to conclude that preferential loss of rhodium from the rhodium-platinum element is unlikely to be a cause of calibration errors at temperatures up to at least 1400°C, although it may need to be considered at higher temperatures. The small amount of direct evidence on this point that is available has been reviewed by Bennett (4), who also describes an experiment in which a complete thermocouple was suspended between terminals and heated by an electric current to about 1500°C for 30 hours in free air. Although the couple lost 16.5% in weight, the calibration at the gold point showed a drop of only 3°C.

c. Migration of Rhodium to Platinum Element

The most important cause of the instability or, as some have termed it, "decalibration" of these thermocouples in industrial service is almost certainly migration of rhodium from the rhodium-platinum to the pure platinum element. In nearly all thermocouple assemblies the two wire elements are run back parallel to each other from the junction, are supported with a spacing of a few millimeters from each other by refractory separators, and enclosed in a metal or refractory sheath. Sometimes the outer tube is packed with powdered alumina or magnesia, sometimes the insulators are threaded on one wire only; sometimes the outer tube forms one element of the thermocouple, but whatever the details of construction it is found that after service the platinum element becomes contaminated along its length with detectable amounts of rhodium that can only be derived from the alloy wire.

It is only very recently that the real cause of decalibration has been fully appreciated, and it has been more customary to put the blame on impurities picked up from the refractories. A careful study by Chaussain (5), for instance, which has been widely quoted, describes a long series of tests on couples and on the wires comprising them, heated in various powdered refractories; yet no spectrographic or other tests at all were made for rhodium migration. This possibility was completely ignored.

The most detailed study of rhodium migration so far published is by Freeman, of the Aircraft Nuclear Propulsion Department, General Electric Company, Cincinnati, Ohio (6). As much as 2% of rhodium was found in the pure platinum wire at a point 2 inches back from the junction of a platinum: 10% rhodium-platinum couple that had been in service, packed with powdered alumina inside a 10% rhodium-platinum alloy sheath.

The mechanism by which rhodium is transported is intriguing, although speculation concerning it is still unfortunately hampered by lack of reliable data. It seems most likely that the volatile rhodium oxide is involved in the process; but it has never, for instance, been established whether a refractory packing in a thermocouple assembly does in fact influence the rate of migration. If, as some have suggested, its presence accelerates migration, presumably it is as a source of oxygen ions that play their part in intermediate reactions involving the volatile platinum metal oxides.

d. Contamination by Impurity Elements

Change in calibration through contamination of the junction by metallic impurities is, of course, most serious when bare thermocouples are used—as in experimental work when an unprotected couple is wired to a specimen. It must be recognized that all base metals alloy readily with the platinum metals and diffuse rapidly at elevated temperatures. Contact with molten base metals is obviously undesirable, if only because they all rapidly attack the thermocouple wires, taking the platinum into solution. However, contact with any hot, solid base metals should always be avoided as far as possible since solid solution occurs so very readily. Above about 1000°C, for instance, diffusion of iron into platinum is very rapid, and if it extends beyond the region of constant temperature, may seriously influence the emf generated. Thus it is always desirable to employ such means as a wash of high-grade alumina cement to avoid direct contact between a bare couple and a hot metallic object.

Apart from causing changes in calibration, many impurities may embrittle platinum thermocouple wires and lead to fracture. The commonest embrittling agent is silicon, derived from silica or siliceous refractories by reduction. If platinum is heated in contact with silicon dioxide or silicates in a reducing atmosphere (or even in vacuum) to above about 1100°C, it reacts with the refractory, and free silicon is formed. The reduced silicon immediately alloys with the platinum and diffuses away from the site of the reaction, so that reduction is encouraged to continue. Silicon has a relatively low solid solubility in platinum but forms a platinum-platinum silicide eutectic, which collects at the grain boundaries. This eutectic melts at a temperature well below 800°C, so if the wires are heated above this temperature the boundaries separate under the slightest applied stress and the wires disintegrate.

It is thus essential that silica-containing refractories should be kept from contact with platinum thermocouples unless reducing conditions can be rigorously excluded. Moreover, it is equally important to ensure that conditions in the thermocouple assembly are such as to avoid the formation of traces of the volatile compound SiS_2, which will also embrittle platinum or platinum alloys. SiS_2 is readily formed by the reaction of sulfur with silica or siliceous refractories in the presence of carbon as a catalyst. Such dangerous conditions may arise from the presence of small amounts of many common oils or greases. Traces of lubricating oil have in particular been found to provide sufficient sulfur and carbon to produce damaging amounts of SiS_2 in "quick-immersion" thermocouple assemblies used for measuring the temperature of liquid steel (8)—even in regions where the wires are quite out of contact with silica. The trouble can be avoided by "burning out" the steel tubes at a red heat, blowing air through at the same time, to remove all traces of oil before assembly.

Platinum thermocouples, unlike base metal couples, are not contaminated or embrittled by hydrogen, carbon, or sulfur. In special applications, therefore, where adequate precautions against contamination by reduced oxides can be taken, bare platinum couple wires can be used in contact with very pure graphite or in atmospheres of hydrocarbons, sulfur dioxide or trioxide, sulfuretted hydrogen, or other compounds of these elements.

B. PLATINUM: IRIDIUM-PLATINUM

The only thermocouple in this system that has been used to any important extent is that of platinum with the 10% iridium-platinum alloy. This combination was first selected by Professor G. Tait in 1872, in a study of the thermoelectric phenomenon, but as early as 1886 Le Chatelier recommended the platinum: 10% rhodium-platinum as being more reliable. The 10% iridium-platinum alloy is generally regarded as being susceptible to loss of iridium at high temperature through preferential oxidation. Thus, although the emf generated by coupling it with platinum is about half as much again as that from the platinum: 13% rhodium-platinum couple, it has been very little used.

There are, however, sufficient numbers of recorders and indicators calibrated for this couple still in service to maintain a demand for replacements. Manufacture of the alloy is carried out in the same way as for the rhodium-platinum alloys, and the composition adjusted to keep the emf within ±1°C of standard value at the gold point. A condensed calibration is given in Table VI.

Technically, the couple is in every way as suitable as the platinum: 13% rhodium-platinum couple over the range of temperatures from —50° to at least 850°C, and it may be used with confidence, at any rate for

TABLE VI
TEMPERATURE-EMF RELATIONSHIPS FOR PLATINUM:
10% IRIDIUM-PLATINUM THERMOCOUPLES

Temperatures °C[a]	emf, mV
100	1.34
200	2.82
300	4.40
500	7.66
700	10.98
1000	16.04
1200	19.40
1300	21.07
1500	24.38
1600	26.00

[a] Cold junctions at 0°C.

occasional service, at temperatures up to 1600°C. Prolonged exposure to temperatures in the higher range may, however, as indicated above, cause calibration changes. The same precautions against contamination are necessary as with all platinum couples.

C. RHODIUM-PLATINUM:RHODIUM-PLATINUM

For measuring temperatures in the upper range, particularly above about 1400°C, there are advantages in using rhodium-platinum alloys for both thermocouple elements, even though the emf developed is appreciably less than when one of the elements is pure platinum. Briefly these advantages are

a. The calibration is more stable, being less influenced by migration of small amounts of rhodium.

b. The melting point of the junction is raised.

c. The effect of variations in the temperature of the cold junction is, by reason of an inflection in the calibration curve, greatly reduced.

d. Danger of fracture by creep of the weaker platinum element is reduced.

The rhodium-platinum compositions that have been used are

$$1\%:13\% \text{ Rh}$$
$$5\%:20\%$$
$$6\%:30\%$$
$$20\%:40\%$$

A reference table for the 6% Rh-Pt:30% Rh-Pt couple was issued by the National Bureau of Standards in 1966. Condensed calibrations for this and the other three couples are given in Table VII.

TABLE VII

APPROXIMATE TEMPERATURE-EMF RELATIONSHIPS FOR VARYING COMPOSITIONS OF RHODIUM-PLATINUM: RHODIUM-PLATINUM THERMOCOUPLES

Tempera-ture, °C[a]	emf, mV			
	1% Rh:13% Rh	5% Rh:20% Rh	6% Rh:30% Rh	20% Rh:40% Rh
200	~0.4	0.03	—	—
500	1.5	1.4	—	0.30
1000	5.0	4.9	4.8	1.35
1200	6.8	6.7	6.8	1.75
1300	7.7	7.7	7.9	2.1
1500	9.7	9.6	10.1	3.1
1600	10.9	10.6	11.3	3.6

[a] Cold junctions at 0°C.

The 13% rhodium-platinum alloy wires used in making these combinations are commonly taken from batches made and calibrated against platinum as described in Section II. The remaining alloys are usually made by melting weighed amounts of the two constituent metals and the composition is checked analytically. Rhodium-platinum:rhodium-platinum couples are not usually adjusted to provide calibrations so closely matched to standard tables as are the couples described in Section II,A, and for work of the highest accuracy it is advisable to prepare a special calibration for each batch of wire.

Choice among these couples for any particular applications appears to be largely a matter of personal preference, although the following rather scanty evidence of their characteristics has been gathered.

1. Stability of Calibration

It seems to be tacitly accepted that there is little to choose between these couples in their sensitivity—or, rather, insensitivity—to change of calibration from rhodium migration. Thus, when the main concern is to guard against this cause of inaccuracy—in, for instance, thermocouples controlling the temperature in creep furnaces, which are likely to remain undisturbed for many hundreds or thousands of hours—the preferred couple is usually the 1%:13% combination. This is appreciably less costly than the others and very slightly more sensitive.

2. Melting Point

When a couple combination is selected for service at high temperatures, the melting point of the lower-melting limb may be a governing factor. The solidus temperatures of the rhodium-platinum alloys are approximately as follows:

1% rhodium	1770°C
5% rhodium	1825°C
20% rhodium	1850°C

There is some uncertainty as to the exact values of these solidus temperatures, but the shape of the solidus curve is well established and it seems reasonable to conclude that the solidus temperature of the 40% rhodium alloy is in fact very little higher than that of the 5% alloy.

All in all, the 20:40 rhodium-platinum couple seems to offer only very slight advantages over the 5:20 rhodium-platinum combination. One manufacturer puts forward the following recommendations for the temperature ranges that can be measured with the couples:

| | Maximum temperature, °C, for | |
Rhodium-platinum combination	Continuous service	Spot readings
1%:13%	1400	1650
5%:20%		
6%:30%	1500	1700
20%:40%	1600	1800

In at least one reference in the literature, however, it is stated that the 5:20 couple can be used continuously up to 1700°C and for spot readings up to 1825°C, its melting point.

3. Cold Junction Compensation

With all rhodium-platinum:rhodium-platinum thermocouples, the emf developed at temperatures up to about 100°C is extremely small, with the result that when these couples are used for high-temperature measurement, large variations in the temperature of the "cold junction" can usually be neglected. It should, perhaps, be emphasized that it is necessary to add the *emf* corresponding to the temperature of the cold junction to the instrument reading to correct for the cold-junction error—and not to add the actual temperature of the cold junction to the indicated temperatures.

The corrections to be applied to 5:30 rhodium-platinum couples are compared with those necessary for platinum:13% rhodium-platinum couples.

Cold-junction temperature (°C)	emf correction, mV	
	Pt:13 Rh-Pt	5:20 Rh-Pt
1	0.005	—
2	0.011	—
5	0.027	—
10	0.054	0.001
15	0.083	0.001
20	0.112	0.001
25	0.142	0.002

The error in the readings of 6:30 rhodium-platinum couples at various temperatures, caused by increases in the cold-junction temperature up to 100°C, is shown below.

Cold-junction temperature, °C	Error in recorded temperature, °C, at							
	600	700	800	1000	1200	1400	1600	1800
20	−0.8	−0.7	−0.7	−0.5	−0.5	−0.4	−0.4	−0.4
60	−4.4	−3.7	−3.4	−2.8	−2.5	−2.3	−2.3	−2.3
100	−9.4	−8.0	−7.4	−6.1	−5.3	−5.0	−5.0	−5.0

Since in industrial applications the difficulties of maintaining the cold junction at a constant temperature tend to increase at very high temperatures of operation, the insensitivity of the rhodium-platinum couples to quite large variations in cold-junction temperature is often a very real point to be considered in their favor. It alone may justify their higher cost.

4. Creep Strength of Wires at High Temperature

The increased creep strength of the low-alloy wire as compared with that of pure platinum may be useful in installations in which these wires are subjected to stress—as for instance when each wire is required to support the weight of a string of insulators. The following comparative figures for creep strength at 1450°C have been published:

Material	Stress, psi, to rupture in	
	10 hr	100 hr
Platinum	310	130
10% Rhodium-platinum	940	510

Figures for 5% rhodium-platinum are not available, but the stress for failure is generally considered not very much less than that for 10% alloy.

D. Fibro-Platinum : Rhodium-Platinum

It has been suggested that there are advantages in using the material known as fibro-platinum (9) in place of pure platinum as one element of platinum:rhodium-platinum thermocouples. To make fibro-platinum, a bar of pure platinum ½ inch in diameter is bored out to produce a tube having an inside diameter of ¼ inch with ⅛ inch walls. The tube is then packed with as many pure platinum wires as possible, each wire being a standard thermocouple element 0.020 inch in diameter. The assembly is then reduced by swaging and wire-drawing to a final diameter of 0.020 inch.

This treament produces a material which tends to recrystallize in very long elongated crystals and imparts a fibrous structure to the wire, increasing appreciably its resistance to creep at high temperatures. Little data appear to be available on the stability of the fibrous structure at various temperatures or on the exact effect of the treatment on the purity and emf of the product. At 1450°C, the stresses required to cause failure in 10 hours and 100 hours respectively are quoted as 540 and 370 psi, respectively, as compared with 310 and 130 psi for pure platinum.

E. Chromel : Alumel

Chromel-alumel couples were introduced by Hoskins Manufacturing Company of Detroit nearly half a century ago and are probably the most widely used of thermocouple combinations. The emf developed is relatively high; they may be employed over the whole of the temperature range from 0 K to 1300°C and can be relied on for spot checks up to about 1400°C.

The two base-metal alloys of which they are composed are relatively cheap and remarkably resistant to oxidation when heated in air to temperatures up to about 1300°C (2400°F). Their manufacture is controlled so that the emf developed by the couples is consistent to a high degree from batch to batch. The alloys are susceptible to attack by sulfur and sulfur-bearing gases, and may be contaminated if heated in carbonaceous atmospheres. The emf developed is large and the couples are very stable in service under oxidizing conditions.

1. Chromel P

"Chromel" as a generic name has been applied by Hoskins Company to a series of nickel-chromium alloys, all containing 15 to 20% of chromium and from 35 to 85% of nickel. For thermocouple purposes, the alloy presently supplied is known as chromel P and is basically a 90:10 nickel-chromium alloy. The manufacturers describe it as "a nickel-chromium alloy containing nearly ten times as much nickel as chromium, but also incorporating nine additional minor constituents." Its melting point is described loosely as 1430°C, and its resistivity is about 85×10^{-6} ohm-om at 20°C, with an average temperature coefficient of resistivity from 20° to 100°C of 3.2×10^{-4} per°C.

2. Alumel

Alumel is essentially an impure nickel. The composition is described as comprising 95% of nickel, the balance being silicon, manganese, and aluminum. The maximum temperature at which it should be used is stated to be 1300°C.

3. Calibration and Stability

The emf developed by the chromel-alumel couples is not so high as that from such couples as iron:constantan; but the couple resists oxidation to higher temperatures and is generally regarded as the best all-around combination. Commercial chromel:alumel couples are supplied to comply within ±¾% with the calibration tables issued by the U. S. Bureau of Standards Circular No. 561, from which the values shown in the abbreviated Table VIII have been extracted.

In tests reported by Potts and McElroy (10), it was found that after exposure to air at 1000°C for 500 hours, chromel P wires gained weight slightly by 120 mg/mm² and their calibration at the silver point changed by −1.8%; and that alumel wires gained 300 mg/mm² and their calibration after 65 hours changed by +17%. This change reported in the calibration of the alumel wires is greater than is usually expected.

TABLE VIII

TEMPERATURE-EMF RELATIONSHIPS FOR CHROMEL
P-ALUMEL THERMOCOUPLES

Temperature, °C[a]	emf, mV
−200	−5.75
−100	−3.49
0	0
+100	4.10
200	8.13
300	12.21
500	20.64
700	29.14
1000	41.31
1200	48.89
1400	55.81

[a] Cold junction at 0°C.

The same authors found that wires as installed were often cold-worked sufficiently to cause an error of 3°C at 300°C. Normally, when couples are used at high temperatures, the wires will become annealed before any emf readings are taken. If couples are to be used below 400°C, it is recommended that they should first be heated for 4 hours at 400°C, after which they will remain stable within ±½% at 400°C.

In the absence of oxygen, the wires are normally regarded as being quite stable. However, for the most accurate work, it is advisable to avoid the use of oxide-coated wires and to use only bright-annealed couples. Couples made from oxidized wires may show drifts of up to —3 mV in 150 hours in stagnant helium at 1000°C, probably due to selective oxidation of chromium and reactions of the surface oxides with carbon.

F. Copper:Constantan

The couple between copper and constantan dates from the early days of thermoelectric measurement. It develops a high emf, and is limited in its range of application only by oxidation of the copper in air at temperatures above about 500°C.

High conductivity electrolytic tough-pitch copper as used for electrical conductors is invariably picked for the copper limb, and no special requirements are set for thermocouple applications. Constantan is an alloy of long standing in the electrical industry, originally developed for its low temperature coefficient of resistivity. Its composition is, basically, copper, 60%, nickel, 40%, with small amounts of manganese, silicon, and iron as the principal impurities.

Batches of the alloy may be chosen for thermocouple purposes, but no special attempts are made, as far as is known, to match them to a standard calibration table. The couple is particularly useful for services at low and ambient temperatures. For accurate work, batches are usually

TABLE IX
Approximate Temperature-emf Relationships for
Copper-Constantan Thermocouples

Temperature, °C[a]	emf, mV
—200	—5.54
—100	—3.35
0	0
100	4.28
200	9.29
300	14.86
400	20.87

[a] Cold junction at 0°C.

calibrated individually; the approximate temperature-emf relationships are set out in Table IX.

G. Iron:Constantan

Iron:constantan couples develop an even higher emf than copper: constantan couples. They are a little more rugged mechanically and need only the protection afforded by a tube slipped around the wires to permit them to be used at temperatures up to 1000°C without damaging oxidation.

Any batch of soft (low-carbon) steel wire that is available is often the source of the iron leg of this couple. The iron wire used by florists in the days before plastics was a common source and may still be obtainable.

At low and ambient temperatures, the susceptibility of the iron wires to corrosive attack is a frequent bar to the use of this couple, and at high temperatures the ease of oxidation of the wires is a similar hindrance. Added to these disadvantages is the risk of rusting of wires when not in use. It is easy to overemphasize these hazards, however, and the couple probably deserves more attention than it gets.

Large batches of wire can be made with a high degree of uniformity, and could be calibrated individually. The emf-temperature relationships of the couple are indicated in Table X.

H. Molybdenum:Tungsten

As a practical means of measuring temperature, the molybdenum: tungsten thermocouple has apparently little to recommend it, save only that it will withstand, without melting, temperatures beyond the melting points of any other usable metals. These are temperatures beyond which

TABLE X

Approximate Temperature-emf Relationships for
Iron-Constantan Thermocouples

Temperature, °C[a]	emf, μV
−200	−8.27
−100	−4.82
0	0
100	5.40
200	10.99
300	16.56
500	27.58
700	39.30
1000	58.22

[a] Cold junction at 0°C.

few refractories retain their strength or shape. The melting point of molybdenum is 2620°C. The emf developed is small, and is often unstable. Moreover, the wires oxidize readily even when protected, embrittling seriously in service. Nevertheless, it is still probably the commonest choice if a thermocouple must be used for temperatures in excess of about 1650°C. The upper limit of temperature at which it can be used is determined more by the limitations of refractory insulators than by the melting or softening of the wires.

The molybdenum wires generally used are of the regular powder metallurgy material, bright-annealed below about 1100°C so as to retain their fibrous structure and ductility. Tungsten wires for these couples are similarly taken from regular stock, preferably in long jointless lengths. Very little can be done by way of adjustment to ensure conformity to a standard calibration table. Undoped and unthoriated tungsten is usually employed.

The emf generated by the couple shows the unusual features of rising to a maximum at about 500°C, then decreasing to zero at about 1000°C, changing sign, and increasing in intensity fairly rapidy as the temperature is further raised. A typical calibration is given in Table XI.

The couple is clearly useful only at temperatures above about 1200°C. The possibility of substituting a molybdenum:tungsten alloy for molybdenum in order to eliminate the inflection in the pure metal couple has been investigated, but alloys containing up to 50% of tungsten all behave like molybdenum when coupled with tungsten.

The stability of the couple is often called in question. However, tests

TABLE XI

APPROXIMATE TEMPERATURE-EMF RELATIONSHIPS FOR
MOLYBDENUM-TUNGSTEN THERMOCOUPLES

Temperature, °C[a]	emf, mV
100	0.33
200	0.57
300	0.75
500	0.90
600	0.87
700	0.75
900	0.38
1000	0
1100	−0.39
1200	−0.87
1300	−1.30
1400	−1.65

[a] Cold junction at 0°C.

by Kuether (11) have shown that the tungsten limb at least shows no instability at temperatures up to 3600°F (1980°C) and that some batches may be stable up to 4500°F (2480°C), once initial stresses have been relieved.

The variation of calibration from batch to batch is, however, not negligible. Kuether has reported variations in emf of only ±4 μV at 650°C along the length of one batch of tungsten, but found variations of ±300 μV in six batches of tungsten from the same supplier. It is thus essential to calibrate each batch of couple wires individually.

Two characteristics of tungsten and molybdenum impose limitations on the practical applications of the couple. The first is mechanical—the brittleness that develops when these materials are heated above their recrystallization temperature—say 1200°C. Tungsten wires, in particular, lose all room-temperature ductility after once being heated, and the couples must be left in position, virtually undisturbed, once they have been heated up to temperature.

The second characteristic is chemical—the susceptibility of both metals to oxidation at high temperatures and their affinity for carbon. Molybdenum, in particular, is attacked rapidly in air, and a volatile oxide is formed. The couple normally needs the protection of a reducing atmosphere, such as hydrogen, and must be kept from contact with carbon. It is also suitable for use in vacuum.

None of these limitations, luckily, are a serious drawback to its use as a single-shot rapid-reading immersion thermocouple for such purposes as measuring the temperature of a bath of liquid steel. Both elements can be supplied in the form of the thin wires preferred for this application, and it is possible that this combination, known for so many years, may in this way find a new use for controlling the production of new steels and special high-temperature alloys.

I. TUNGSTEN-RHENIUM : TUNGSTEN

Some of the disadvantages of the molybdenum : tungsten thermocouple are lessened if an alloy of tungsten with 26% of rhenium is substituted for molybdenum. The calibration curve shows no inflection, and the emf developed increases uniformly with temperature; it is more than ten times higher than for the molybdenum-tungsten couple at 1500°C. The melting point of the alloy is over 2980°C, as compared with 2620°C for molybdenum. However, the disadvantage remains that the tungsten limb is brittle at room temperature once it has been heated above the recrystallization temperature. This characteristic can be modified to some extent by adding a few percent of rhenium to the tungsten limb.

It has been known for some years that the addition of about 10% or more of rhenium to tungsten is able to suppress the characteristic inter-crystalline brittleness of the refractory metal and imparts ductility. The 26% rhenium alloy was later selected as a thermocouple material on account of its high thermoelectric power against tungsten. The alloy can be made by powder metallurgy or by electron-beam or arc melting, followed by hot-forging and hot-swaging; it is reduced to the final dimensions by wire-drawing.

The couple needs to be operated in vacuum, hydrogen, or an inert gas, and is subject to rapid oxidation in air at high temperatures. As is true for all couples, when it is operated above about 1800°C, the lack of suitable refractories for insulating and supporting the wires makes permanent installations difficult to provide. Little has been published on the variations in calibration to be expected between various batches of wire, and standard tables have not been set up. The approximate relationships between emf and temperature are given in Table XII.

To modify the disadvantageous brittleness of pure tungsten, an improvement has been to add rhenium to both elements of the thermocouple. Thus, a 26% rhenium-tungsten alloy has been coupled with a 5% rhenium-tungsten alloy. Also, a 25% rhenium-tungsten element has been used with a 3% rhenium-tungsten alloy. Such thermocouples can be used at temperatures up to 2760°C (5000°F), but accuracy above about 2300°C may not be guaranteed.

TABLE XII

Approximate Temperature-emf Relationships for Tungsten-26% Rhenium:Tungsten Thermocouples

Temperature, °C[a]	emf, mV
10	0.027
20	0.054
100	0.356
200	1.033
500	4.848
700	8.395
1000	14.527
1200	18.847
1500	24.918
1700	28.569
2000	33.835
2200	37.145
2300	38.325
2320	38.516

[a] Cold junction at 0°C.

J. Iridium:40% Iridium-60% Rhodium

In 1933, Feussner suggested a combination of iridium and 40% iridium-60% rhodium for measuring temperatures above the melting point of platinum, and for this purpose the couple has received token endorsement ever since.

Pure iridium is not easily formed into wire. At the time the couple was introduced, the only method was to hot-swage (by hand) strips cut from sheet that had been hot-forged from powder compacts. Iridium wire is now hot-swaged and sometimes hot-drawn from powder compacts or arc-melted ingots, the practice following closely that used for the production of tungsten. Like tungsten, it is ductile only as long as it retains a fibrous structure and it develops room-temperature brittleness once it has been heated to a point above its recrystallization temperature.

The 40% Ir-60% Rh alloy is almost as difficult to work into wire as pure iridium. Originally, it was probably melted with an oxy-hydrogen torch on a lime block and formed into wire by laboriously hot-forging the small ingots. Today, when wire is made, it is either argon-arc melted or swaged, like tungsten, from powder compacts.

This couple is more resistant to oxidation than the tungsten-molybdenum couple, and indeed, the protection afforded by a closely fitting sheath is adequate for most industrial installations. It is, like all platinum couples, susceptible to contamination by silicon reduced from refractories, and when it cools to room temperature after service the iridium wires, in particular, are brittle and fragile. The emf developed is, as will be seen from Table XIII, about half that developed by a platinum:13% rhodium-platinum couple over corresponding temperature ranges.

TABLE XIII

Approximate Temperature-emf Relationships for Iridium:40% Iridium-60% Rhodium Thermocouples

Temperature, °C[a]	emf, mV
100	0.365
200	0.818
500	2.464
700	3.628
1000	5.308
1200	6.388
1500	8.013
1700	9.156
1800	9.745
1900	10.355
2000	10.995
2100	11.654

[a] Cold junction at 0°C.

K. SILICON CARBIDE:GRAPHITE

For measuring very high temperatures, it is natural to consider the possibility of using nonmetallic materials, and of these particularly silicon carbide and graphite have been studied.

The construction generally favored has been to couple a graphite rod to a silicon carbide nosepiece, which is plugged into one end of a silicon carbide tube surrounding the graphite rod. The space between the rod and the surrounding tube may be packed with alumina, provided the couple is not used at temperatures above about 1800°C. At higher temperatures, alumina attacks graphite, and the space must be filled with an inert gas or with graphite black, which helps to retard thinning of the graphite through oxidation. The emf developed varies appreciably from batch to batch of graphite and of silicon carbide. Also, it is affected by the impurities present, notably by the boron content of the graphite. In general, it may be expected to amount to 150 to 200 $\mu V/°C$ over the whole of the temperature range.

The mass of the thermocouple is an obvious limitation, restricting its use to the measurement of the temperature of rather large sources of high heat content. It is reasonably satisfactory in permanent installations and may be expected to have a life of over 100 hours at 1750°C in carbonaceous atmospheres. In air, the life tends to be restricted by oxidation at temperatures much over 500°C.

L. PALLADIUM-GOLD:IRIDIUM-PLATINUM (PALLADOR)

Even with modern recording instruments, the platinum:platinum-rhodium thermocouples are not always easy to use for measuring temperatures in the ambient range or below, since the values of emf are so low in relation to possible parasitic voltages. The pallador thermocouple represents one attempt to provide an essentially noble metal thermocouple—resistant to oxidation at moderate temperatures and to corrosion—with a relatively high emf. Instead of platinum, an alloy of 40% palladium with gold is used, and this is coupled with 10% iridium-platinum to provide as high an emf as possible.

This 40% palladium-60% gold alloy is normally made by high-frequency melting in a graphite, zircon, or thoria crucible, with precautions to avoid solution of oxygen. The resulting ingot is worked down like any other gold alloy. It is annealed in hydrogen, then quenched to prevent the formation of a tarnish film of palladium oxide on cooling.

The couple is particularly suited for use at temperatures up to about 750°C. Over this temperature range it is stable, although the emf developed tends to vary somewhat from batch to batch of the palladium-gold alloy wire. The approximate values are shown in Table XIV.

TABLE XIV

APPROXIMATE TEMPERATURE-EMF RELATIONSHIPS FOR
40% PALLADIUM-GOLD:13% IRIDIUM-PLATINUM
(PALLADOR) THERMOCOUPLES

Temperature, °C[a]	emf, mV
100	4.5
200	10.0
300	16.0
500	28.3
700	40.8
800	47.1

[a] Cold junction at 0°C.

Above about 900°C, a curious form of instability that is not fully understood is often encountered. The emf, instead of falling as in the deterioration of most thermocouples, is found to rise progressively for some hours. After 15 hours at 1000°C, for instance, a couple heated in air was found to be reading about 19°C high, and on further heating the wires embrittled. It has been suggested that the effect may be associated with the oxidation of the palladium and that it may depend on the original gas content of the alloy; but the mechanism remains obscure. It has been claimed that signficant instability may be overcome by a "stabilizing treatment" or "pre-anneal" in manufacture, prior to drawing the material to wire.

The couple is particularly useful in circumstances where a corrosion-resistant thermocouple is required for use at moderate temperatures—as for instance in biological studies.

M. PLATINEL

The platinel thermocouple is, essentially, a development of the gold-palladium:iridium-platinum couple, designed to match as closely as possible in its emf-temperature relationships the chromel-alumel couple. Its noble metal characteristics make it particularly suited for biological applications, not only at ambient temperatures, but also at temperatures up to at least 1100°C in atmospheres containing steam and CO_2, which attack chromel-alumel couples. The alloys employed are positive element: palladium 83%, platinum 14%, gold 3%; negative element: gold 65%, palladium 35%. The emf developed by the couple is given as 48.0 ±0.2 mV at 1200°C, as compared with 48.89 mV for chromel-alumel at the same temperature.

For many industrial applications, platinel couples may thus be used without serious error with recording and indicating instruments calibrated for chromel-alumel.

At 1000°C the stability of the couple in air is such that no change is to be expected after at least 360 hours. At 1200°C, even in an atmosphere of wet steam and CO_2, the change in calibration after more than 1000 hours has been found to be less than 0.2%. Chromel-alumel couples are completely embrittled after 300–400 hours at 1100°C in this atmosphere.

N. Platinum:Molybdenum-Platinum and Molybdenum-Platinum: Molybdenum-Platinum

These thermocouple combinations have been developed specifically for measuring temperatures in atomic piles, where they are subjected to neutron bombardment.

Most thermocouples contain at least one element that may be activated in these conditions. Molybdenum, however, has the low neutron-absorption cross section of only 2.5 barns and is relatively unaffected. Three combinations have been proposed, of which the first is the most widely used:

1. 0.1% molybdenum-platinum:5% molybdenum platinum
2. platinum:5% molybdenum platinum
3. 1% molybdenum-platinum:5% molybdenum platinum

The alloys are invariably made by induction melting. Owing to the ease of oxidation of molybdenum, it is not easy to control composition within very close limits and the couples are usually supplied with batch calibrations.

The emf developed by the couples is not high, but they may be used with confidence up to at least 1500°C. The alloys are very susceptible, however, to loss of molybdenum through selection oxidation to the volatile molybdenum trioxide. Care is needed to protect the couple, either by surrounding it with a reducing atmosphere or by working in vacuum.

O. Cobalt-Gold:Copper

The cobalt-gold:copper thermocouple is useful in measuring very low temperatures, where the emf developed is sufficiently large for it to be sensitive to temperatures as low as 4°K.

The composition adopted for the cobalt-gold member is pure gold containing 2.1 atomic percent of cobalt. It is usually supplied as wire about 0.005 inch in diameter, severely hard-drawn. To date, only small quantities have been made commercially, and no details of impurity levels, desirable or undesirable, are available. The wires may be covered with a woven Fiberglas insulation, unvarnished.

The copper wires used by cryogenic laboratories in making these couples appear to have been of run-of-the-mill copper as used for electrical purposes—presumably drawn from high-conductivity electrolytic tough pitch copper. They have apparently been annealed and often have been Teflon-covered.

Although this couple is sensitive enough to be one of the most attractive for use at low temperatures, it has been reported (7) to exhibit appreciable inhomogeneities. Short-range differences of up to 4 mV were found in regions separated by only a few inches at temperatures up to 76°K, and differences of up to 14 μV at intervals of a few feet in the range from 76°K to room temperature. Moreover, variations of up to 0.5% were found from batch to batch. The investigators at the Bureau of Standards Cryogenic Laboratory at Boulder, Colorado, found that the calibration was not changed significantly by annealing at 70°C for 24 hours, but that at 90°C the thermoelectric power decreased by about 1% after 24 hours.

However, the tests were all made on completed couples and the possibility of modifying the annealing and cold-working procedures in the production of the wires was not investigated. There seems every likelihood that with further study, it should be possible to devise a means for producing uniform and stable wires.

An extract from a typical calibration is given in Table XV.

TABLE XV

APPROXIMATE TEMPERATURE-EMF RELATIONSHIPS
FOR 2.1 ATOMIC PERCENT
COBALT-GOLD:COPPER THERMOCOUPLE

Temperature, K	emf, μV
1	0.53
2	2.09
3	4.66
5	8.22
10	12.74
20	179.6
30	372.5
50	893.9
100	2622.6
200	6730.6
300	11025.5

P. Cobalt-Gold:Normal Silver

The cobalt-gold:normal silver thermocouple is very similar in its characteristics to the cobalt-gold:copper couple, but it has the slight advantage that since the thermal conductivity of normal silver at low temperatures is slightly less than that of copper, there is rather less chance of error through conduction along the wire.

Normal silver is the name given to an alloy of silver with 0.37 atomic percent of gold. It is used as wire, about 0.005 inch in diameter, severely cold-drawn, and insulated with woven Fiberglas like the cobalt-gold alloy elements. The cobalt-gold alloy for this couple is the same as that used for the cobalt-gold:copper thermocouple.

The inhomogeneities in these couples, as supplied up to the present, appear to be of much the same order as those in the cobalt-gold:copper combination (7). The emf developed is only slightly less, the difference amounting to about 0.01 mV at 10°K and 0.2 mV at 20°K. At higher temperatures the difference is more noticeable, becoming about 122 mV at 200°K. The couple is often preferred for measuring very low temperatures, below 80°K, where heat leaks along the wires become troublesome.

III. Resistance Thermometer Elements

A. Thermopure Platinum

Platinum resistance thermometers are used in industry for a range of temperatures from −183°C or lower to at least 650°C. When suitable precautions are taken to avoid contamination from the supporting framework they have been used up to 1000° or 1100°C. It is important that after winding, the platinum elements should be annealed carefully by heating to above the maximum working temperature, and the windings and their supports must be so designed that the wire remains unstrained at all temperatures.

The production of platinum of the highest possible purity in the form of wire for resistance thermometer elements is carried out under carefully controlled conditions by all the leading suppliers. As a measure of purity, it is customary to rely upon the average temperature coefficient of resistance between 0° and 100°C expressed as the ratio of the resistivity at 100° and 0°C, R_{100}/R_0. The 1968 revision of the International Temperature Scale specifies that this should have a minimum value of 1.3925 in order that the shape of the calibration curve from −183° to 630°C shall be correct.

In general, there is no difference between the wires used for thermo-

couples and for resistance elements, and similar precautions are taken
with both to avoid contamination during annealing and wire-drawing. A
cross section of a platinum resistance thermometer element is shown in
Fig. 2.

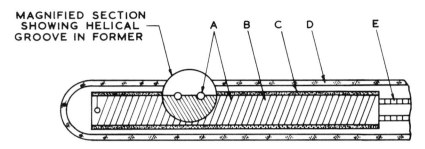

FIG. 2. Typical resistance thermometer element. A, Thermopure platinum wire;
B, former; C, packing; D, sheath; E, insulating bead.

B. PURE NICKEL

For industrial temperature measurements, pure nickel has the outstanding advantage that its temperature coefficient of resistivity in the
range 0°–100°C is higher than that of any other metal available commercially in the form of fine wire. Pure nickel is widely used for measuring
temperatures below about 400°C; at higher temperatures there are
dangers of oxidation. It is desirable to choose a source of high-purity
nickel wire for this application, since the coefficient will be appreciably
higher than that of commercial malleable nickel wire, deoxidized with
manganese and magnesium. Wire drawn from powder-metallurgy bars
pressed from carbonyl nickel powder or nickel powder reduced from
chemically purified oxide may have a value of $R_{100}/R_0 = 1.68$, whereas
for commercial nickel the value may be 1.5 or less.

The resistance elements are commonly wound from hard-drawn
nickel wire insulated with Fiberglas and annealed for some time at
300°C.

C. GERMANIUM

Small bridges of arsenic-doped germanium have been used as resistance thermometers for measuring very low temperatures from 1.5 to
15 K. A typical bridge is about 0.5 cm long with cross section about
0.05 cm square, cut with its length in the (110) crystal direction. Connections are made to gold-plated arms by 0.01% antimony-gold alloy
wires.

The thermometers are stable on recycling and are extremely sensitive, especially below about 10 K. A typical calibration follows:

Temperature, K	Resistance, ohm
1.5	2300
2	789
4.2	101
10	18.3
15	10.5
20	8.0

D. Miscellaneous Temperature-Resistance Elements

Numerous materials have been used as resistance elements for temperature measurements from time to time, but few have received wide application. Some of these are listed briefly below:

Material	Temperature range	Notes
Thermistors (composition not disclosed)	4.2–200 K	Very high temperature coefficient
Alumina	Above 2000°C	Experimental
Indium	5–100 K	
Lead	Below about 20 K	
Gold + 0.15% silver	Below about 100 K	

IV. Expansion Elements

It is seldom that the expansion of one element alone can be used as a measure of temperature; more often a second element is needed as a basis of reference, and the relative expansion of the two is the variable that is observed. When one of the elements is a liquid, as in mercury or alcohol thermometers, the container serves as the reference body. When both the elements are solid, it is important to recognize that constraints to their free movement may influence the relative expansion observed and that any plastic flow resulting from applied stresses will be a cause of instability and drift.

A. Mercury in Glass

This is not the place in which to elaborate on the applications of mercury in glass thermometers, the speed of their response, or the methods of correcting their readings for depth of immersion of the bulb or the length and temperature of the emergent stem. It will be sufficient to note that they can be used for temperature measurement over the range from just above the freezing point of mercury, —38.37°C, to as high as 600°C if a suitable glass is employed and the stem is filled with dry nitrogen under pressure.

The essential requirement for the mercury is that it shall not wet the glass, that its surface shall be clean and bright, and that it shall not leave trails behind as it moves along the capillary. The mercury must accordingly be of high purity, or at least be free from metallic impurities that are liable to oxidize, and from grease and organic matter. In the past, acid treatments and filtration through chamois leather were among the means used to achieve a satisfactory product. Today, vacuum distillation is the usual means of purification. The distilled mercury should be stored in clean glass containers; polythene and similar plastics are liable to cause contamination both from the plastic materials and from the fillers.

Considerable attention has been given for a century or more to the production of glass that will remain stable under thermal cycling and be suitable for thermometer tubing. In recent years development has been directed especially towards glasses stable to higher and higher temperatures. These glasses must not only resist flow under the internal pressure when hot but must be free from secular aging. The following tabulation describes the glasses most widely used.

Glass	Exposure limits, °C		
	Strain point	Continuous	Intermittent
Corning Normal 7560	500	370	430
Kimble R6	490	360	420
Jena 16 III	495	365	425
Corning borosilicate 2954	548	420	480
Corning 1720	668	540	600
Jena Supremax 2955	665	535	595

Fused quartz has also been used successfully for special mercury thermometers. Considerable skill is needed for manufacture of such thermometers, but they are perfectly stable up to 500°C. It is important to avoid contact of hot quartz with alkali-containing refractories, such as fire clay, since traces of alkali may cause surface devitrification.

B. THALLIUM-MERCURY IN GLASS

To extend the range of service of mercury in glass-type thermometers downwards below —38°C, the thallium-mercury eutectic, containing 8.7% of thallium, may be substituted for mercury. The eutectic melts at

—59°C and extends the range accordingly. Thallium, however, oxidizes very readily in air, and special techniques are needed to produce a satisfactory thermometer.

C. GALLIUM IN QUARTZ

Gallium melts at 29.8°C, but its vapor pressure is very low and its boiling point at atmospheric pressure has been reported to be as high as 2250°C. Its use in a quartz envelope for measuring temperatures up to 1000°C was described as long ago as 1926, but very few instruments are believed to have been made.

It would seem likely that gallium may attack quartz and be contaminated after quite short periods of service.

D. ORGANIC LIQUIDS IN GLASS

The most widely used means of extending the serviceability of liquid-filled thermometers to low temperatures is to substitute an organic liquid, usually colored by a trace of dye, for mercury.

For this purpose alcohol, toluene, and pentane are most commonly employed. Commercial pentane, indeed, is said to be serviceable down to —196°C, below the freezing point of the pure product.

For high temperatures, above 150°C, organic phosphates and organo-polysiloxanes have been used, but little data are available on their characteristics.

V. Other Materials Related to Temperature Measurement

A. BIMETALLIC STRIPS

For operation over relatively narrow ranges of temperature, bimetallic strips are used as temperature-sensitive elements in many industrial and domestic devices. Usually, these are actuating devices rather than strictly means of temperature measurement, but they are too important to be omitted from a discussion of the subject.

The strip elements are formed from duplex sheet made by bonding two alloys having widely different thermal expansion coefficients, such as invar (the 70:30 nickel-iron alloy) and 60:40 brass. Slabs of the two alloys are scalped or ground on their faces, butted together, and hot-rolled with a heavy initial reduction. The slabs may be welded at their edges and are sometimes enclosed in a thin nickel or iron envelope to exclude oxygen during heating. The composite is then usually cold-rolled to the finished size.

Strips cut from this material flex one way or the other as their temperature is raised or lowered due to the differences in expansion of the two components. If one end is clamped, the movement of the free end may be used to indicate the temperature of the strip or to open or close contacts at a predetermined temperature.

The curvature assumed by the strip is the resultant of a complex system of forces and is influenced not only by the expansion character- istics of the two alloys but even more by the relative thickness of the two layers, by the total thickness of the strip, and by its width and length. The elastic constants of the two alloys are also factors. The force that can be exerted by the free end as the result of a given temperature change is a matter of concern in many applications and needs design consideration. The choice of alloy is often less important than good engineering in the successful utilization of a bimetallic strip.

The most widely used combination is that of invar with 60:40 brass. This is reasonably stable at temperatures up to 150°C and is suitable for thermostats for such domestic items as irons, water heaters, and electric blankets.

At higher temperatures, plastic deformation at the interface (always present to some extent) introduces serious hysteresis, so that the strip becomes unreliable as a temperature indicator. Other combinations, more stable in these circumstances, make use of stainless steels with nickel-chromium alloys, and have been employed to a limited extent.

B. TEMPERATURE SENTINELS

Although not strictly materials for temperature measurement, various "temperature sentinels" warrant at least a brief mention. These take several forms but all are installed to give visible indication—either during a heating cycle or after it has been completed—of when or whether a predetermined temperature has been attained.

1. Pyrometric Cones

"Seger cones" were introduced in Germany in 1886 and "Orton cones" were developed by Dr. Orton of Ohio State University in 1896. These pyrometer cones have since been standardized by the American Society for Testing and Materials and are in the form of slender pyramids, pressed usually from mixtures of china clay, feldspar, whitening, and flint, molded to size and carefully dried. Two sizes are available: small cones 1⅛ inch long with ⁹⁄₃₂-inch base and large cones 2⅜ inch long with ⅝-inch base. Sets of these, softening at a series of temperatures, are introduced with the work in the kiln, and arranged on a refractory plate inclined at 8°. When the softening temperature is reached, the apex of the pyramid bends over; and by examining the cones after firing, the observer can form an estimate of the maximum temperature of firing. The batches of refractory from which the cones are made are carefully

standardized for oxidizing conditions in the kiln; but the softening point
depends not only on the temperature but also on the rate of heating. For
this reason they may in some instances be a better measure of the effects
of a firing cycle on a batch of ceramic ware than the readings of a thermometer of the maximum temperature alone, but this does not make
them any more reliable as temperature-measuring materials. About fifty
different cones are available, with softening points varying in steps of
about 20°C, as shown in Table XVI.

2. Alloy Sentinels

These consist of short lengths of wire that can provide an indication
that their melting point has been exceeded. They have been particularly
valuable for monitoring the temperatures reached at various points in
moving pistons of internal combustion engines, where thermocouples

TABLE XVI
PYROMETRIC CONES

Series	Cone number	Softening temperature, °C	
		Heated at 20°C per hour	Heated at 150°C per hour
Soft series (12 cones)	022	585	605
	to		
	011	875	905
Low-temperature series (10 cones)	010	890	895
	to		
	01	1110	1145
Intermediate temperature series	1	1125	1160
	to		
(20 cones)	2	1135	1170
	to		
	10	1260	1305
	to		
	20	1520	1530
		Heated at 100°C per hour	Heated at 600°C per hour
High-temperature series (20 cones)	23	1580	—
	to		
	38	1835	—
	to		
	39		1065
	to		
	42		2015

could not be affixed. The wires are preferably made of pure metals or eutectic alloys and will tend to shrink and become globular in form almost immediately on reaching a temperature above the melting point.

3. *Heat Fuses*

Often fuses are installed so as to protect furnace windings from overheating. These are also constructed from a range of pure metals or eutectic alloys. They are wired in series with the heating current or separately in an alarm circuit, and melt when their melting point is exceeded. A necessary precaution is to mount them under slight tension so that oxide skin will be fractured as soon as the wire melts; otherwise the molten core may remain intact and continue to conduct.

C. Thermosensitive Paints and Crayons

Brief mention may be made of the many proprietary mixtures available for application as surface temperature indicators. These are generally based on colored salts of cobalt, copper, nickel, or manganese, which are dispersed, often with oxide extenders, in conventional paint mediums or waxes. When applied to bearing housings, furnace bodies, or similar surfaces they provide a visual indication that a given temperature is exceeded.

The changes in color on heating are seldom clean-cut, and unless the heating rate is very slow, may occur over a range of temperature. Moreover, since the color changes are often associated with the loss of combined water, they are not usually reversible with temperature.

The characteristics of some typical paints are given in Table XVII.

TABLE XVII
TYPICAL THERMOSENSITIVE PAINTS

	Heating period		
Principal constituent	10 minutes at	120 minutes at	Color change
Ferrite yellow	300°C	260°C	Yellow to red brown
$CuSo_4 \cdot 3\ Cu(OH)_2 \cdot H_2O$	230	195	Green to brown
$NiCl_2 \cdot 2\ C_6H_{12}N_4 \cdot 2\ H_2O$	120	100	Yellow to violet
$CoCl_2 \cdot 2\ C_6H_{12}N_4 \cdot 10\ H_2O$	40	33	Red to blue

D. Compensating Leads for Thermocouples

For reasons of economy, it is sometimes desirable not to use thermocouples sufficiently long to extend to the cold junction, but to terminate them at a terminal block close to the furnace shell. To avoid error due to unknown differences in temperature between the terminal block and the cold junction, "compensating leads" may be used to make the connection.

The basic requirement for these leads is that the emf generated by their junction should match the emf of the thermocouple over the limited range of temperature involved—say from 0°C to 100°C. The wires do not need to withstand oxidation at high temperatures, can be selected from the cheapest materials available to match all the common thermocouple combinations, and are conveniently supplied insulated and braided. The earliest compensating leads were in the form of multistrand conductors twisted from combinations of thin copper and nickel wires. One conductor, for instance, might consist of eight copper and two nickel wires, whereas the other consisted of one copper and nine nickel wires. By changing the combinations, it was possible to match either chromel: alumel or platinum:rhodium-platinum couples.

Today, most compensating leads employ copper as one element and a copper-rich copper-nickel alloy as the other. By varying the nickel content in the range from about 2 to about 15, compensating leads can be provided for all the usual thermocouple combinations.

References

1. T. Wensel, Temperature. *In* "Temperature," Vol. 1, p. 19. Reinhold, New York, 1941.
2. E. H. McLaren, The freezing points of high purity metals as precision standards. *In* "Temperature," Vol. 3, (C. M. Herzfeld and A. I. Dahl, eds.), Pt. II, pp. 185–197. Reinhold, New York, 1962.
3. W. Betteridge, D. W. Rhys, and D. F. Withers, Laboratory control of production platinum for thermometry. *In* "Temperature," Vol. 3, (C. M. Herzfeld and A. I. Dahl, eds.), Pt. 1, pp. 263–267. Reinhold, New York, 1962.
4. H. E. Bennett, "Noble Metal Thermocouples," 3rd ed. Johnson Matthey, England, 1961.
5. M. Chaussain, Platinum-platinum/rhodium thermocouples and their industrial applications. *Proc. Inst. Brit. Foundrymen* **44**, A60-77 (1951).
6. R. J. Freeman, Thermoelectric stability of platinum vs. platinum-rhodium thermocouples. *In* "Temperature," Vol. 3 (C. M. Herzfeld and A. I. Dahl, eds.), Pt. II, pp. 201–220. Reinhold, New York, 1962.
7. R. L. Powell, L. P. Cawood, Jr., and M. D. Bunch, Low-temperature thermocouples. *In* "Temperature," Vol. 3 (C. M. Herzfeld and A. I. Dahl, eds.), Pt. II, pp. 65–77. Reinhold, New York, 1962.
8. Liquid Steel Temperature Sub-Committee Symposium. *J. Iron Steel Inst.* **155**, 213 (1947).
9. J. S. Hill, Fibro platinum for thermocouple elements. *In* "Temperature," Vol. 3 (C. M. Herzfeld and A. I. Dahl, eds.), Pt. II, pp. 157–160. Reinhold, New York, 1962.

10. F. J. Potts and D. L. McElroy, The effects of cold working, heat treatment, and oxidation on the thermal emf of nickel-base thermoelements. *In* "Temperature," Vol. 3 (C. M. Herzfeld and A. I. Dahl, eds.), Pt. II, pp. 243–264. Reinhold, New York, 1962.

11. F. W. Kuether, Measurement of high temperature thermal emf characteristics. *In* "Temperature," Vol. 3 (C. M. Herzfeld and A. I. Dahl, eds.), Pt. II, pp. 229–232. Reinhold, New York, 1962.

THERMAL INSULATION SYSTEMS

E. C. Shuman

Professional Engineer, State College, Pennsylvania

I. Introduction

A. Nature of Thermal Insulation

Thermal insulation is often thought of as merely a type of material because literature and advertising of thermal insulating materials for homes, buildings, and industrial constructions appear in all news media. Aeronautics and space travel publicity also has impressed upon the public that thermal insulation is an essential part of systems in which heat flow occurs. For a given exposure, heat flow rates depend upon the resistance to heat flow of complete constructions. Consequently, the thermal insulation industry has been urging designers and users to think in terms of thermal insulation systems rather than thermal insulation materials.

Committee C-16 of Thermal and Cryogenic Insulating Materials of the American Society for Testing and Materials defines thermal insulation as a material or an assembly of materials used primarily to provide resistance to heat flow (1). The materials used may not be in themselves good insulators. Assemblies of materials could provide spaces which may be filled with liquids, air or other gases, or may be partially or almost totally evacuated so that the materials serve only as boundaries. A thermal insulation system is a combination of thermal insulating materials and means for attachment to or with the surfaces to be insulated and with facings, vapor barriers, and protective coverings *as installed*. Figure 1 shows typical complexity of industrial piping systems, and illustrates the need in such systems for insulations with physical properties suited for different kinds of attachments and sometimes for allowed slippage in service. The insulation, of rigid type, was preformed into hemicylindrical sections 3 feet long, a general industry standard, and is being attached with steel wires. Subsequently, the insulation will be protected from weather by jacketing or coating (to be described later).

Thermal insulation is a misunderstood class of material, except to the relatively few who understand the principles which affect performance of thermal insulation systems. Insulation systems have widely differing applications, many of which are highly complex and include costly materials having special properties, so that it is impossible to delineate a thermal insulation industry. For example, refractories function as thermal insulations for temperatures above about 1800F but are usually considered a separate class of material. Refractories and "standard" thermal insulations are often combined in thermal insulation systems. The scope of the industry cannot be defined, but annual sales of thermal insulation in the United States are greater than a quarter billion dollars.

Since performance of the thermal insulation portion of a whole construction is of concern, this chapter will present the principles of heat flow as they apply to performance under the environments anticipated by the designer; the effects of temperatures within the system on moisture migrations, many of which are undesirable and have often been overlooked in the past; and the need to consider several properties of a given material, not thermal resistance alone.

Acceptable performance of thermal insulation systems is the responsibility not only of the producer of materials, but of the designer, the

FIG. 1. An industrial pipe insulation application. The importance of the property of handleability in insulation materials is evident. Courtesy GAF Corporation.

erecting contractor, and the owner-operator. Marketable materials are engineering compromises of properties and capabilities. Thermal insulations are in use from absolute zero temperature range to those so high that they deteriorate the material and must be evaluated in systems according to expected exposures and required duration of service as matters of engineering operations and economics. Since engineering properties govern design of systems, obviously there is no "best" insulation because properties other than the often used k factor may be overriding. While thermal insulation systems can do no more than the properties of the total system permit, proper maintenance of the systems by the user is expected.

Thermal insulation systems for simple structures, as in the walls, floors, or roofs of a house, are more complex than many suppose, and ignoring the applicable simple laws of physics has led to more perform- ance problems than is reasonable. To point to performance of thermal insulation systems rather than merely to some properties of specific types of materials, this chapter will be treated virtually without mathematical expressions. The approach will be application of simple principles that apply to all systems and will apply to any new materials not yet on the market. In addition, examples of inadequate performance in common constructions will illustrate the significance of thermal insulation systems as opposed to thermal insulation materials. The materials mentioned will be those in common use in constructions and processes, rather than those that have been developed for limited applications, such as the suits for astronauts costing $45,000, or materials with high thermal resistance and limited thickness, costing $10 or more per square foot, justifiable as such costs may be for the performances required. The economics of the thermal insulation systems may be complex, but they should be studied carefully; for example, cryogenic fluids may be best transported through pipes at high flow rates for short durations uninsulated, even outdoors in direct sunshine, despite the large temperature difference from ambient to fluid.

The important properties that need evaluation for different applica- tions will be discussed, with some examples of troubles in performance that have occurred when these auxiliary properties have not been con- sidered. As for all materials, there can be no true substitution of insula- tions; although one material may be used in place of another, some properties will be lost while others may be gained. The way the change in material affects the performance of the insulation system must be determined from an engineering standpoint, with respect to installed costs over the expected service life.

Consideration of moisture effects must be made simultaneously with evaluation of the thermal system behavior. The need for dual analyses is so great that ideally both should be discussed in separate columns on the same page. Examples are given to illustrate operating problems blamed upon insulation material that were in fact due to lack of protec- tion from moisture.

Moisture in this discussion means water in its three forms, liquid, solid, or vapor. The problems due to moisture in insulation systems are not dependent on the state or combination of states in which it is present; the word will therefore be used to imply whatever state or combination is present in a specific location in the insulation system. A multistate water system involves total pressure differences, but the water vapor state is, fortunately, the most frequent state encountered in constructions. Water vapor is gaseous and is commonly considered a component of air, but air may or may not contain moisture. An appreciation that the behavior of moisture is independent of air is important.

Since products from different producers of the same classification of insulation may vary in their properties, the information in later sections of this chapter on general evaluations should not be used alone for selection of specific materials. Most of the insulations on the market are described in ASTM specifications (1). However, those specifications describe minimum values so that materials which meet them will give acceptable performance in usual environments, but specific brands may exceed one or more of the basic requirements significantly. The need for the more important properties in a particular application is determined by the designer of the thermal insulation system.

The principles of physics that apply to thermal insulations, and for that matter, to all constructions, are few and simply stated. The simplicity of these principles seems to have led to disdain for their validity. The examples of errors given in the last section are intended to reemphasize the importance of these laws in all constructions, but data for specific designs are beyond the scope of this discussion.

B. Value of Thermal Insulation Systems

Several reasons for using thermal insulation systems are: if wanted heat flows out of a construction, the lost heat is an expense; if unwanted heat flows into a construction, the removal is an expense; if either situation induces an undesired change in moisture within the construction, correction of the condition is expensive but may lead to performance savings, whereas noncorrection entails increasing operating expense and premature failures of equipment.

The designs of insulation systems are predicated upon the economics of applied insulations, rather than on the cost of materials, and insulated systems are compared with uninsulated or inadequately insulated systems. A difficulty today in rational analyses of ultimate cost is estimation of the value of heat in the future, but despite this difficulty an estimate should be made for the life expectancy of the application. In some applications in which few, if any, changes in plant are expected, a highly durable and thermally efficient insulation system may be indicated. In other plants which characteristically change arrangements of equipment or remove insulations periodically for inspections of lines or equipments, a compromise is developed by which lesser efficiency is accepted to gain ease of removal, thereby minimizing reinstallation costs.

The general bases for using thermal insulation systems are

1. Heat Flow Limitation.
   a. For conservation of heat.
   b. For exclusion of heat.
   c. For process temperature maintenance.
2. Moisture Migration Limitation.
   a. To maintain environments internal and external to spaces.
   b. To maintain process requirements.
   c. To avoid deteriorating condensations.
3. Protection.
   a. Against personal injury from burns.
   b. Against damage to adjacent constructions with temperature or heat interchange limitations.
   c. Against fire. Thermal insulation systems designed primarily as fire protection, usually misnamed fireproofing, as though fires can be rendered ineffective, are beyond the scope of this paper, but, frequently, fire hazard potentials associated with thermal insulations are a major concern.
   d. Against radiation. Thermal insulations may be used as part of a system designed to protect life and property from heat radiation, but may themselves be evaluated for their susceptibility to high-velocity radiation.

C. Historical Development

Although present-day materials are the primary concern, a brief history will be helpful. New materials were developed as the need arose. The basic principles and much of the mathematics of heat flow were published in 1822 by J. B. J. Fourier, and by Laplace, Poisson, Peclet, Lord Kelvin, Riemann, and others, but very little practical use was made of the information until about 1900. Temperatures encountered in processes in use up to then were comparatively low, so that organic materials such as cork, vegetable fibers, kelp, shavings, or hulls of nuts and seeds provided sufficient thermal resistance. As higher surface temperatures began to be used, inorganic materials with high thermal resistance were sought, and about 1885 a synthetic product called magnesia, later called 85% magnesia, was invented. This was suited to the high temperatures of the time, about 500F. Later, when high-pressure steam and chemical processes were developed with operating temperatures of 1000F or more, diatomaceous earth with incorporated asbestos fibers was molded into blocks and tubular segments, called pipe insulation. The higher temperature material was used to reduce the temperatures to below 500F so that 85% magnesia could be applied over it to reduce the temperature of the exposed surfaces to the limit desired, usually below 140F so that personnel would not be burned on contact. The use of two different materials is an example of engineering economics; the diatomaceous material did not provide as much thermal resistance per inch of thickness as 85% magnesia, and cost more, so only enough was used to provide thermal resistance to reduce interface temperature below the operating temperature limit of the magnesia.

In the early 1940's, two companies not previously in thermal insulation business introduced, in substantial volume, hydrous calcium silicate insulation in flat block and curved sections for pipes and vessels. Up to an operating temperature of 1200F, there was no need for two materials for service on most equipment in use at that time and today. Since the thermal resistance per inch of thickness of calcium silicate is essentially the same as that of 85% magnesia, the sale of 85% magnesia has dwindled to a negligible volume. The former large producers of 85% magnesia have converted to hydrous calcium silicates. However, because some equipment operates at above 1200F, there is still sale of diatomaceous earth insulation, some of which is suited to temperatures of 2000F. One of the calcium silicate producers developed a particular type that is suitable up to about 1700F.

Fibrous wool-like materials formed from molten rock, slag, or glass but of differing temperature limits have been available for several decades, but have been improved in recent years.

In 1920, the National Bureau of Standards published a resume of thermal conductivities of a long list of materials available at that time. (The resume is now out of print.) For hot services, only two or three of the materials listed then are still in use in substantial sales volume.

Present-day thermal insulations serve widely differing markets, from house and building construction and transportation media to complex chemical processes. Insulations that have appeared in recent years and have remained in production have provided a combination of useful physical properties in addition to acceptable thermal properties at competitive or advantageous costs installed. Properties other than thermal resistance, or the old style k factor, govern selection of types of insulation for specific applications as was shown in Fig. 1.

The large markets are in building construction, particularly houses, and the industrial process and power complexes. The exotic insulations developed for applications in which high thermal resistance at minimal weight were essential, regardless of cost, perform under the same physical laws as other common insulations, and despite their excellent performance in these special applications, they are not suited to the major markets.

D. Types of Systems

Thermal insulation systems may be generally classified along environmental or functional lines. Thermal insulations, whether used as an

operational necessity or to effect economies, require engineering evaluations peculiar to each installation. However, the basic principles remain
the same. Past experience in seemingly simple constructions, such as
houses, indicates that overlooking the simultaneous effects of heat and
moisture has led to unnecessary troubles. The same is true, in fact, for
all insulation systems. No sharp demarcations can be drawn between
systems; the laws of physics apply to all.

The most important types of insulation systems are the following:

1. Equipment Systems

Most large vessels, pumps, and tanks are parts of industrial operations, and are usually outdoors, so that compatible insulations and
enclosing envelopes must be selected.

2. Piping Systems Above Ground

Most large pipe lines and associated valves and fittings are in industrial operations, and many of them are outdoors so that compatible
insulations and enclosing envelopes must be selected. Underground
piping systems need special designs that are often complex and difficult
to keep dry.

Special designs are needed for heat-traced piping, a system by which
auxiliary heat is supplied all along the line when normal operation is
interrupted and the material in the line must not be allowed to cool
below a critical temperature (see Fig. 10).

3. Underground Distribution Systems

Although underground systems are a necessity in some industrial
operations, they are avoided insofar as practicable. A major problem is
flooding by accident and damage which cannot be repaired without
major costs. Moreover, it is often essential that flow in piping not be
interrupted even though flooding has occurred. Large underground piping systems are used in multibuilding constructions, not only in cities
but in outlying hospitals, institutions, and residential complexes. In some
heavy snow regions, underground piping systems are made as passageways between buildings. In these cases, the thermal insulations should be
adequate to limit heating of passageways to comfortable temperatures.

4. Cryogenic Systems

The term *cryogenic,* which refers to the very low temperature range,
is no longer defined as strictly as formerly for liquid oxygen range, but
usually means colder than -100 to $-150F$ (-75 to $-100C$). The
insulation design problems due to moisture ingress in cryogenic systems

are not found in the same magnitude in higher temperature systems. An excellent survey of cryogenic systems was published in 1967 by the National Aeronautics and Space Administration (12).

5. Building Insulation Systems

Houses and apartment buildings are a very large market for insulation in walls, under roofs and floors, and there are special needs for perimeter insulation, sill insulation, and prevention of moisture migration. Commercial and high-rise offices and apartments are a growing market, especially since glass areas and complete air conditioning are becoming a necessity.

Roof insulations, on level roofs especially, develop troubles too frequently because of moisture ingress in conjunction with diurnal and seasonal temperature cycles, particularly when buildings are air conditioned continuously for high relative humidities, as in textile mills. Too often, roof insulations become wet because motions of structural members are so great that the roofing cracks because it can not withstand such strains, especially after aging.

6. Cold Storage and Freezer Systems

The need for economic installation for heat flow limitation in cold storage and freezer systems is obvious. In addition major problems arise with moisture ingress into walls and roofs, and the formation of ice beneath floors on ground causing substantial heaving of the floor with damage to the building components. Such problems require not only special designs, but also specially experienced contractors and personnel to install the insulation system. In food storage locations, a frequent health safety requirement is an interior finish that permits disinfecting periodically.

7. Transportation Systems

Aerospace devices and vehicles require special thermal insulation systems with unusual developments of components not generally needed in the normal applications of insulation. Nevertheless, those systems perform under the same laws of physics for heat and moisture flow as any other insulation system. An example is the fogging of the astronaut's face mask because he faced cold outer space instead of the gold-lined end of the capsule.

Other modes of transportation of people and materials also require thermal insulation systems designed for particular environments, often where abuse and vibration are common, as on trains and buses.

II. Basic Principles of System Design

A. Physical Laws of Heat and Moisture Flow

The reader is presumed to be familiar with the very significant difference between heat and temperature.

The simple natural laws that apply to all constructions, and must be considered in all applications of thermal insulations are these:

1. Heat flows only from a higher temperature toward a lower temperature.
2. Water vapor flows only from a higher vapor pressure toward a lower vapor pressure.
3. For acceptable performance, stresses must not exceed safe working limits under the service exposures.

Heat is transferred, or flows, due to a temperature difference in only three modes, separately or in combination. These modes are described here for review. Their relative significance will be discussed as they apply to specific materials or performance problems.

1. Conduction—transfer of heat through solid substances.
2. Convection—transfer of heat through nonsolid substances such as fluids; for most thermal insulations the substance is primarily, but not exclusively, air.
3. Radiation—transfer of heat through space, which may be partially or totally evacuated.

In design of thermal insulation systems with respect to the installed size and efficient performance of associated equipment, the rate of heat exchange (the heat loads) for several specific exposures must be evaluated. Two primary exposure types are:

1. Steady State Heat Flows

Unvarying temperature differences that create heat flows which continue for long periods (days or longer) are considered to be steady state even though some small variations occur. When a process plant is started for operation, rates of heat flow through insulation are often not considered until operating conditions become established. However, in some parts of the process, rates of heat flow that affect process temperatures may require evaluation of the thermal diffusivity of the material as well as its steady state thermal resistance.

2. Transient Heat Flows

Temperatures that vary continuously, or continue only for short periods (hours or less) induce transient heat flows, that depend upon the thermal diffusivity of the material. Heat flow induced by transient conditions such as unpredictable temperature excursions, is not determined by diffusivity alone, but requires design consideration of the property of maximum-use temperature of the insulation, to be discussed later.

Since constructions are designed for the maximum heat loads anticipated, usually with some factor of safety, resort to transient state analyses of the insulation is seldom necessary; such analyses are highly complex.

B. Thermal Properties of Insulation Materials

Although this discussion is essentially nonmathematical, some terms, their symbols, units, and relationships should be understood, because they are used to describe how different materials and combinations perform in retarding heat flow under differing conditions.

Historically, the rate of heat flow through one square foot of material has been described as its thermal conductance, and materials have been compared on the basis of the conductance per unit thickness, usually one inch, which then is called *conductivity,* or in engineering terms, *k* factor. Although there is no technical error in comparing materials by *k* factor, it is virtually useless to the nontechnical consumer for evaluating the performance of a material as installed. The home owner or home builder may see thermal insulating materials being installed, but even when the products are marked with *k* factors he cannot tell whether or not the installed materials will provide the expected resistance to heat flow.

For some years there has been a growing appreciation that for nonmetallic materials, and thermal insulation in particular, a more realistic property is *thermal resistance, R.*

1. Thermal Resistance, R

The total resistance to heat flow along the path of temperature difference is usually expressed in British thermal units (Btu) per square foot per hour for an insulated surface, or in SI units as watts per meter, degrees Kelvin or Celsius. For pipes and curved sections, calculations involve an intermediate area known as the logarithmic mean area, but in manufacturers' literature the values are usually translated into heat loss or gain per linear foot of pipe for convenience. The analogy of flow of heat and flow of electricity, which is usually expressed $I = E/R$, is

$$\text{Heat flow} = \frac{\text{temperature difference in F}}{R}$$

Btu per square foot per hour.

R provides a simple comparison between materials or between constructions by indicating the temperature difference across a construction, such
as a wall, required to drive one Btu through one square foot per hour.
Hence, for unit area and unit time R is merely a temperature difference
per Btu. If homogeneous material is one inch thick, then its R is the
reciprocal of conduc*tivity*, or k factor, the Btu's that one degree of temperature difference will drive through one inch per hour. For nonhomo-
geneous constructions—and all installed insulation systems are nonhomo-
geneous—it is simpler and less conducive to errors to consider only
thermal resistance, because in a series of materials and spaces the
resistances may be added, whereas k factors may not.

Terms ending in *ance* indicate the factor for the thickness or the
construction as used, whereas terms ending in *ivity* indicate a property of
a homogeneous material per unit thickness, usually one inch, but in some
cases one foot.

The range of thermal resistivity of some common insulating materials
is shown in Fig. 2.

When several materials are in series, one over the other, and the
areas are large so that edge effects are negligible, the calculations with
k factor for flat surfaces and steady state take the form

$$\text{Heat flow} = \frac{\text{area} \times (\text{temp. hot face} - \text{temp. cold face})}{(\text{thickness}_1/k_1) + (\text{thickness}_2/k_2) + (\text{thickness}_3/k_3) + \cdots}$$

For calculations with thermal resistances, the form is simpler:

$$\text{Heat flow} = \frac{\text{area} \times (\text{temp. hot face} - \text{temp. cold face})}{R_1 + R_2 + R_3 + \cdots}$$

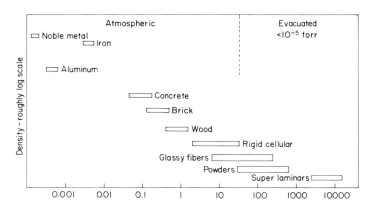

FIG. 2. Thermal resistivity of materials in insulation systems.

For multilayer pipe insulations, the calculations are much more complex, because logarithmic mean areas for each layer enter into the equation.

For materials with high conductances, such as metals, units of thermal conductivity may be preferable for calculations by ordinary means; for electronic computation, either unit may be used.

a. Evaluation of Thermal Resistance

As mentioned earlier, and repeated here for emphasis, thermal conductance numerically is the reciprocal of thermal resistance. Also, as we have seen, when a series of materials are combined so that heat flows through all of them, the thermal resistance of the series is the sum of the individual resistances. On the other hand, thermal conductances may not be added. The result would be erroneous.

In evaluating heat flow, many users of thermal insulations in constructions consider only the areas in which heat flows through parallel thicknesses, but the heat flows through all of the construction in these areas perpendicularly to the faces, regardless of shape. Although the faces are parallel, this movement should not be confused with parallel heat flow, but is series heat flow, through one material or space after another. In constructions in which the insulation is placed between the studs, heat flows through the studs in a direction parallel with flow through the insulated areas but at a different rate. This is parallel heat flow. At the intersections of studs and insulation, the heat paths are highly complex, in accordance with the physical law that heat flows from higher temperature to lower regardless of the direction. Evaluations would have to be made at every spot in minute fractions of an inch. It is virtually impracticable to calculate accurately the rates of heat flow for complex combinations of series and parallel heat flows; it is difficult even for sophisticated electronic computers. When well-insulated areas are large, the parallel heat flow, as through studs, nails, metal attachments, etc., is considered a heat leak. The designer must decide whether constructions that combine different shapes and different materials induce heat leaks of magnitudes sufficient for estimation of their effects. In an ordinary house, these stud heat leaks may be several percent of the total heat loss, but in general, the effects of air infiltration through poorly fitting windows and doors, and of family habits of leaving doors open are much greater than the heat leaks of studs.

For many years, the guarded hot plate has been the standard method for heat flow measurements for flat or "board" insulations about one inch thick. Another standard method is used for conventional tubular pipe insulations, but most of these are limited to about a 3- or 4-inch pipe diameter. For practical purposes, heat transfer rates through conventional pipe insulations can be calculated if the thermal resistivity is known. Heat flow measurement techniques to obtain reliable results are highly specialized, and a recent publication (2) in two volumes presents a most comprehensive discussion of procedures and pitfalls in heat flow measurements.

Precise determinations require that steady heat flow rates be established before measurements are acceptable, and this usually requires several hours. The guarded hot plate is still considered to be the referee method for thermal insulations, but some simpler and much more rapid methods are acceptable for any needs, especially for rapid checks of quality control in production. One of the rapid methods is known as the "heat flow meter" and the values obtained are acceptable under materials specifications if the reference material being used has been calibrated as prescribed.

Most published values of thermal resistivity or of thermal conductivity are for 75F mean temperature, although values are available for other temperatures from producers and in some data guides (3). An example of the effect of mean temperature on heat transfer rate for one type of fibrous material of one particular fiber characteristic is shown in Fig. 3. Some materials show relatively small effects of large mean temperature variations and some rare ones reverse the effect.

Thermal resistivity is a property of a homogeneous material. However, if the specimen is a combination of materials which together perform as though homogeneous, then thermal resistivity is an acceptable designation. A specimen is considered homogeneous when the resistivity is

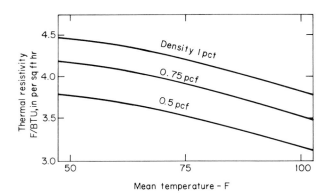

Fig. 3. Effect of mean temperature on the resistivity of a typical mineral fibrous insulation.

unaffected by variations in thickness and area within the range normally used.

In evaluation of a more complex combination of materials, such as a typical wall of studs, interior and exterior facings with or without spaces and with or without emissive or reflective surfaces, then the test apparatus is the "guarded hot box." Hot boxes of various sizes are in use, but generally the portion of the specimen called test area is four feet square or more. Larger areas are preferable but are expensive to provide. One laboratory uses an 8 by 10 foot test area on specimens up to 20 feet wide and 16 feet high. Relatively large test areas are needed to provide facilities for typical wall or panel constructions with enough studs, braces, or other details so that the heat flow rate obtained will be representative of the actual structure. Since orientation affects the rate of heat flow, some hot boxes have been built to be rotable so that heat flow may be measured up, down, sloping, or horizontal.

In the past, values of thermal conductivity (k factor) have been published to the third decimal, such as 0.167, which implies that 0.166 is too small and that 0.168 is too large. None of these values is realistic. Except by accident, three separate specimens selected according to a statistical sampling plan would not agree so closely. During field application of insulation, any joints between sections, especially those which occur when the extremes in dimensional tolerances are adjacent, would affect heat transmission to a greater extent than differences in the third-place decimal. In insulations attached by metal nails or metal studs, these attachments are heat leaks; a 0.1 inch diameter nail through one square foot of a good insulation would increase the heat loss in zero weather by 5 to 7%. When through-metal attachments are most practicable, as is often true for some types of industrial insulations, a compromise between fastening dependability and heat loss due to "leaks" is necessary. In cold service, however, the designer would have to consider that the head of a nail through insulation might well cause condensation of moisture on the head and promote rusting, wetting of the insulation, or other deteriorating effects.

More realistic use of heat transfer data becomes simple for the non-expert if thermal resistance, R, is used. For the k factor 0.167, R = 6, so that if normal manufacturing variations of 0.1 are average, R of 5.9 and 6.1 would be equivalent to k of 0.170 and 0.164, respectively. This difference of about 3% is the agreement expected between different tests of the same material. In view of normal manufacturing variations, use of the apparently but not actually precise values shown above is less than realistic. In process in the ASTM committee on thermal and cryogenic insulation, a current tentative specification for blanket insulation such as that used in houses, requires that resistance R of the material in the package be marked on it conspicuously, preferably in whole numbers but never to more than one decimal.

Although there may be use for some highly precise values of thermal conductance in specialized applications using metal or other high conductors, this order of precision is not usual for the insulation industry.

This discussion shows that it is necessary to consider complete insulations systems of materials, attachments, coverings, open spaces, reflectances, and through-metal or other heat leaks in heat transfer mechanisms. A factor not yet discussed is wind and weather. Heat loss (or gain) from a wall, window, roof, pipe lines, process equipment wall, or any heat transmitting surface outdoors is affected by wind because still air provides some thermal resistance that wind "blows away" or at least reduces. However, on well-insulated systems, the effect of wind on heat transfer is very small, and it is usually either neglected or assigned as an arbitrary wind effect of 7.5 or 15 miles per hour, as illustrated in Table I. Note that the example is for "steady-state" conditions. For transient conditions of diurnal variations of sun and weather, the calculations are so complex that electronic computers are still a necessity.

Omitted from Table I were the studs or lateral bracing between which the insulation is installed and which are heat leaks because their resistances are lower than that of the insulation. Actually, a correction for stud effect in a wall can be made (3). However, such an adjustment is usually not made because weather, frequent openings of doors, mechanical ventilation, laundry and washing, combustion air in furnaces and other variables are so different for different occupancies that such adjusted calculations may be meaningless, especially when the cost of house insulation is considered in terms of the perceptual yearly savings to the home owner if increased thermal resistance is installed.

Table I illustrates that in even moderately insulated walls the principal resistance is due by far to the insulation. Consequently, designers of walls for houses and ordinary buildings can well consider only the R of the insulation for any exposure, and specify the R for minimal, moderate, or superior performance. If house insulation is marked only with its R at standard 75F mean temperature, the owner or designer can decide at the construction site whether or not the insulation delivered was intended to provide the thermal resistance expected. Of course, it is expected by the producer of the materials that the design and construction, including the normal maintenance by the owner, will keep the insulation dry.

In calculations of the example in Table I, numerical values are carried to two and sometimes three decimals, not because these indicate precision of the physical properties but merely to aid in arithmetical checks and

TABLE I

THERMAL RESISTANCE OF INSULATION SYSTEM FOR ONE TYPE OF HOUSE WALL
WITH WIND EFFECT, COMPARED WITH THAT OF UNINSULATED CONSTRUCTION

Construction element	Thermal resistance, R F/Btu (sq ft, hr)	
	No wind	15 mph wind
Indoor air film (still air)	0.68	0.68
Plaster wall	0.40	0.40
Air space (approx. 1 inch)	0.97	0.97
Thermal insulation (approx. 2½ in.)	10.00	10.00
Narrow air space	—	—
Sheathing with building paper	1.04	1.04
Siding	0.85	0.85
Outdoor air film	0.68	0.17
Total R	14.62	14.11
Difference	0.51	

Decrease in resistance due to wind:

$$\frac{0.51}{14.62} = 3.5\%$$

For the same walls without insulation:

	Total R	4.62	4.11

Decrease in resistance due to wind:

$$\frac{0.51}{4.62} = 11\%$$

Heat flow per sq ft for zero outdoors, 75F indoors:

Insulated walls $\quad \dfrac{75}{14.62} = 4.78 \qquad \dfrac{75}{14.11} = 4.95$ Btu/hr

Uninsulated walls $\quad \dfrac{75}{4.62} = 16.1 \qquad \dfrac{75}{4.11} = 18.1$

balances, especially when temperatures within multi-interface constructions are evaluated. The sum of the temperature drops from hot environment to cold must equal total temperature difference from inside to outside. Note again that this analysis applies only to steady-state heat flow.

Such analyses for a house or building illustrate that thermal resistance as a matter of total design is essential as a part of the decision on the size of heating or cooling equipment to specify. Maximum expected heat flows dictate all of the mechanical, power, fuel, and instrumentation needs for operators to accomplish the aims of the installation. Equipment and boilers could be made large enough to waste the heat from uninsulated lines and still have enough available to operate the processes. However, the excess boiler capacity to allow for wasting heat is in reality a charge against the uninsulated system. In buildings which are air-conditioned (note that there are at least four factors of air-conditioning: temperature, relative humidity, circulation, and filtering or cleaning), the question of how large to design the equipment has several aspects. Shall the heating or cooling facilities be large enough for all probable weather exposures, or can temperatures be a bit uncontrollable for a few days out of a year? Equipment varies in capacity in steps, not by infinitely small increments. If the heat loads suggest an equipment size between two available sizes, shall the smaller be specified because it reduces first-cost and investment charges, even though it may very occasionally be undersize? Or shall the larger size be specified because there is good probability that increased capacity will be needed in the future? Shall decisions as to the resistance requirements be predicated upon present costs or shall choices be made in the direction of greater resistances because prices keep on rising?

b. Thermal Resistance and Moisture Migration

Precaution dictates that thermal resistance evaluations be accompanied by virtually simultaneous consideration of the very important factor of moisture migration. Standard moisture pressures vary with temperature, as shown in Fig. 4, and the magnitude of pressure that does occur in wet roofing, for example, should be noted—it is high enough to support several men per square foot. Temperature differences induce vapor pressure differentials, but the relationships are complex (3), and tables or graphs are needed to determine dew points (condensation potentials). The thermal values in Table I show that insulated walls induce warmer interior surfaces, but colder exterior surfaces. The colder outer temperatures aggravate moisture migration tendency appreciably. The potential deterioration due to moisture within walls (or other components) is avoidable by a properly designed vapor barrier system. Vapor barriers, which in reality are vapor flow retarders rather than absolute barriers, are obtainable in many sheet materials, and for buildings should have permeance rating of less than one perm and often less than 0.5 perm. Placement of vapor barriers, if needed, depends on climatic environments. In temperate zones, the vapor barrier would probably be needed on or near the warm or interior surface for winter service. In hot climates, the barrier might be needed on the outer side. In refrigerators the barrier would need to be very much better than 0.5 perm and would definitely be placed on the outside. For freezers, perm ratings of 0.01 or lower may be needed. In some localities where long periods of high relative humidities in addition to warm temperatures are encountered, no vapor barrier or one of high perm rating may be the better design. In

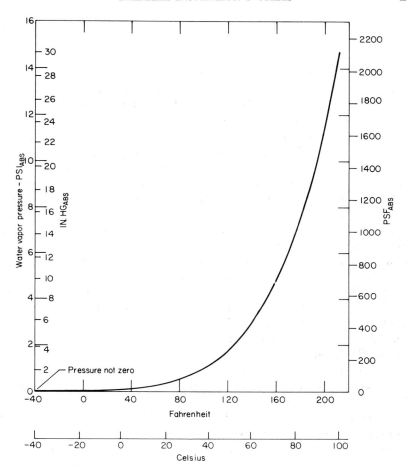

Fɪɢ. 4. Effect of temperature on water vapor pressures in the range of many thermal insulation systems.

all cases, vapor barriers must be free of cuts and holes or discontinuities of any kind, caused either accidentally or for some seemingly good but actually false excuse.

Performance of any construction, insulated or uninsulated, may be based on analyses similar to that above. Thermal performance will probably be more readily understandable in the future than in the past when only thermal resistance units will be used for evaluating designs and installations to avoid unwanted heat transfer.

The recommendation in brief is to insulate on the inside to speed heating or cooling of the space enclosed by outer walls and to insulate on the outside if it is also desired to minimize thermal stresses in the construction.

2. Thermal Diffusivity

When transient states or temperature variations on one or both sides of a construction need evaluation, all the heat that enters one surface does not flow all the way to the other surface. Some of the heat is absorbed by the intervening materials, including any entrapped moisture, which may thus undergo change of state from liquid to vapor. Therefore, the rate of temperature rise on the face opposite that through which heat enters depends not only upon the thermal resistance, but also upon the heat capacity of the materials in the construction. Heat capacity is expressed in British units as the density of the material (pounds per cubic foot, pcf) multiplied by the specific heat (Btu per pound per degree Fahrenheit). The property of material expressed by the ratio of the thermal conductivity to the heat capacity is called *thermal diffusivity*. If all units of the complex ratio are simplified, the resulting unit is square feet per hour, which seems anomalous, but it is published in some literature because it is a necessary factor in certain applications. Although the mathematical expression reduces to square feet per hour, its physical significance is thermal conductivity per unit density and unit specific heat for that material.

The property of diffusivity is important in transient heat flows.

A specific example of transient heat flow occurs in fire exposures, where a sudden temperature rise is created on the outside of a construction. Fire protection is afforded by materials that remain in place and transmit heat very slowly to the construction beneath. Materials that transmit heat slowly, useful for fires of short duration, such as a few minutes, have high heat capacity, that is, have relatively high unit weight and specific heat so as to absorb heat before passing some of it inward. Obviously, heavy-density insulations do not have as high thermal resistance as many light materials, but the high thermal resistances published for insulation usually assume steady-state heat flow, not transient, and a long time (hours) is required for some thicknesses to attain the steady-state temperatures. Serious errors have been made by using so-called highly efficient thermal insulations on the steel legs which support towers, on the supposition that the "better k factor" would provide better fire protection than the heavier, less thermally efficient materials; the light-density materials with low heat capacity which were used failed in a few minutes so that the steel legs became overheated, they crumpled and the tower fell. Heavier materials with their greater heat capacity would have delayed the temperature rise of the steel so that there might have been some opportunity for fire-fighters to extinguish the blaze before enough heat reached the steel to weaken it. For complete evaluation of materials for severe fires, other factors such as changes of state or composition must be included in addition to thermal resistance, R.

However, in ovens in which rapid temperature rise is a process need, light density materials, those with low thermal diffusivity, are needed because they do not absorb much heat so that the oven temperature rises more quickly. In some processes, it is important that ovens cool quickly, and here, again, the low heat capacity of the insulation is a requirement.

In process equipment, insulation is sometimes needed within the vessel or pipes when heat transfer rates between sections of the process system require control. If the operation involves transient variations, either with fluid in contact with the insulation or not, the interposed insulation may need to possess either low or high thermal diffusivity in addition to satisfactory working temperature limit.

3. Thermal Reflectance

When radiant heat strikes a surface, some heat is absorbed and some is reflected. The portion reflected depends upon the reflective property of the exposed surface of the material. The ultimate property for best polished surface, independent of shape, is reflectivity; for the performance of materials as used, under effects of usual surface treatments and cleanliness, the preferred term is reflectance.

Most materials, including metals, unless the surface is highly polished, reflect relatively little heat. Polished gold does not tarnish easily and is a good reflector, but its cost except in special applications is prohibitive. The best of the common materials with low cost is polished aluminum, which may reflect, when clean, 95% or more of the radiant heat striking it. Energy from the sun is radiated over a broad spectrum of frequencies, or wavelengths. The frequency of the visible light portion is appreciably higher than the heat portion, but the heat portion still includes a rather broad band of frequencies. Consequently, the reflectance of materials for the frequency to be encountered must be known. Although commercially polished aluminum reflectors are highly efficient for reflecting long-wave heat energy, they are only about half as efficient when used as reflectors of short-wave heat energy. However, they can be used to reflect visible light. The high reflectance of cleaned polished aluminum in the long-wave heat range makes it suited for resisting energy which emerges from a construction such as the bottom surfaces of a roof or within walls. Aluminum foil placed about an inch beneath the roof surfaces reduces appreciably the sun heat that would otherwise reach the interior ceiling. For best performance in buildings, a space of about one

inch of free air should be provided *uniformly* across the reflecting surfaces. Nonuniform spaces induce higher convection which reduces the
total resistance of the construction. If a free air space is provided across
both surfaces of the foil, even greater resistance to heat flow will be
provided. The effect of temperature differences on rate of heat transfer is
great, since the rate varies as the fourth power of the absolute temperatures $(460 + \text{temp } F)$, as shown in the simplified relation

$$\text{Btu} = \epsilon\sigma(T_a{}^4 - T_b{}^4) \text{ per sq ft, hr}$$

in which ϵ is thermal emissivity and σ is the Stefan-Boltzmann constant.
The numerical value of emissivity is unity minus reflectivity $(1 - \text{refl.})$,
and indicates that the portion of radiant heat striking a surface that is not
absorbed is reflected. This simplified mathematical relationship shows
that the lower the emittance, the less heat progresses through the construction; emissivity is a basic property of a material and emittance is the
performance of the material as installed (discussed again later). Some
effects are illustrated in Figs. 5 and 6.

4. A Recapitulation

Thermal insulations are produced from materials that provide resistance to heat flow in one or more of the three modes. Selection of materials
for specific installations represents a compromise among the properties

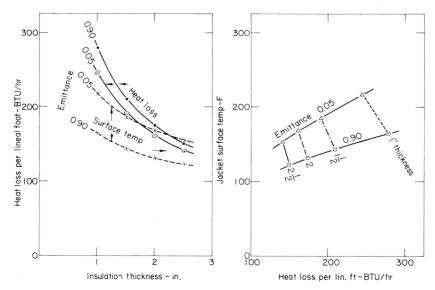

FIG. 5. Effect of emittance of covering over pipe insulation on hot surfaces on
heat loss and exposed surface temperatures, at varying thicknesses of insulation.

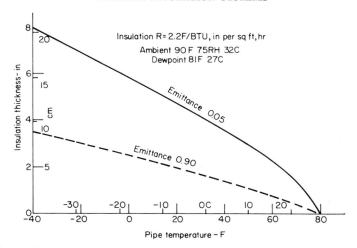

FIG. 6. Effect of emittance of covering over pipe insulation on cold surfaces on required thickness of insulation to keep jacket surface above dew point. Based on calculations by Dr. H. E. Parmer.

of the materials so that the temperature range may be served economically. When a material of relatively low conductance, as a solid, is combined with a material of high reflectance, and the construction minimizes convection, the combination should provide high thermal resistance. This was the aim of the so-called superinsulations in which very small low-conductance fibers were placed between thin sheets of highly reflective films so that there were as few points of contact as practicable, and air was evacuated to a pressure of less than 10^{-5} Torr. Evacuated multilayer reflective assemblies, in which 50 to 100 or more layers per inch of thickness were used, created thermal resistances many times greater than still air, which at one time was considered to be the ultimate. Some insulations have been produced which provide resistance greater than still air but are not as complex as the superinsulations. However, all are far too expensive for the major markets of thermal insulations, and do not represent any new principles. Any development of new materials will be predicated upon the principles already discussed.

C. ECONOMIC CONSIDERATIONS

The cost law of diminishing returns applies to thermal insulation systems; the first thin layer of insulation is the most effective in reducing heat transfer. Subsequent thicknesses in usual applications reduce heat flow farther but at additional cost for insulation, coatings or coverings, adjunct materials and labor. Against these costs, which are primarily first costs, although maintenance costs cannot be ignored, the value of heat not lost may be credited. The designer must decide how much insulation to specify for the best return on the investment for the installed insulation.

The most widely accepted economic analysis for insulations was published by L. B. McMillan (4) in 1927, and the mathematical values for specific designs were derived through the principle which he described (Fig. 7). Since not all operations are continuous, it is usual to analyze them on a yearly basis. The most economic thickness is determined from the minimum of the total cost of fixed charges of insulation installed and the cost of the heat that is still being lost. Obviously, the unit cost of heat and the potential future increase in unit cost at a specific locality is the major factor for determining the type and thickness of insulation to use. Since local costs vary, it is usual for multiplant widespread operations to specify different types and thicknesses in different plants of the same company. As mentioned above, maintenance, inspection removals, and the service conditions that depend upon the number of sizes of insulation kept in stock are best served by compromises. A material which is highly efficient for some portion of an operation but less efficient in other portions may be used for both, so that the number of types and sizes to be kept in stock may be a minimum.

A danger in selecting insulations and the insulation coverings for a specific design by merely considering purchase costs in random lists of both types of materials is the possible incompatibility of materials to be combined and the effect of reflectance (or emittance) of the covering. Adverse effects due to overlooking the need for evaluation of reflectance (or emittance) of coverings were illustrated in Figs. 5 and 6.

The economics of insulation systems in large process operations involves not only decisions as to thickness of insulation of a particular type,

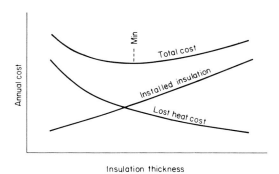

FIG. 7. Relation of incremental cost of additional thickness of insulation to the resultant heat savings and total cost. After L. B. McMillan (4).

but evaluation of the effect that greater thermal efficiency might have on selection of the size of equipment. Cost comparisons should be made on equipment sizes not only as planned for immediate operations, but also for expected expansions. A major computer-calculated extension of the mathematically tedious analyses of McMillan was undertaken in 1960 by Union Carbide Corporation, South Charleston, West Virginia, in co-operation with the University of West Virginia. Variables included were applicable to such a wide range of cost factors that it aroused much interest in the industrial insulation industry. It was adopted by the National Insulation Manufacturers Association, who published it in graphical and tabular form to aid in selection of thicknesses of pipe insulation and flat block for above ambient temperatures (5). Typical results from earlier studies by Union Carbide Corporation are given in Table II.

Two significant factors should be noted. One is the great heat saving achieved by the first layer of insulation. (Hypothetically, thinner layers would show a saving also, but problems of handleability in this size would make them impractical.) The second is the diminishing return as the thickness increases.

The following tabulation provides an example of one means to aid in selecting an economic thickness.

Insulation thickness, inches	Yearly cost of insulation per lineal foot
1.5	$2.55
2.0	2.36
2.5	2.37
3.0	2.45

Apparently, the most economic thickness is about 2 inches. The theoretical low point in cost of insulation might be slightly greater or slightly less than 2 inches, but commercial thicknesses increase only in nominal half-inch steps. Therefore, cost studies must be made on the actual thicknesses to be used, and these may be either slightly more or slightly less than the designated, or nominal thickness. The designer would decide whether the lesser or the greater thickness is preferable. In general, with rising cost of operation ahead, when the difference in installed cost is small, the greater thickness is usually selected.

In an attempt to develop a similar economic analysis for cold lines and flat surfaces, it was found that for many installations, the economic thickness was insufficient to avoid condensation, so that designs to minimize moisture problems could not be based upon thermal analysis alone.

TABLE II

ILLUSTRATIVE COST ANALYSIS OF PIPE INSULATION[a]

Insulation thickness, inch	Capital investment to provide steam for heat loss	Capital investment in insulation	Total capital investment	Yearly cost of heat lost	Added cost per ½ inch of insulation	Fuel saving per ½ inch of insulation	Percent return
(bare)	$28.32	none	$28.32	$18.13	—	—	—
1.5	2.57	$3.30	5.87	1.64	—	—	—
2.0	2.10	4.10	6.20	1.34	$0.33	$0.30	90
2.5	1.74	4.84	6.58	1.14	0.38	0.20	53
3.0	1.52	5.90	7.42	0.98	0.84	0.16	19

[a] Data from Union Carbide Corporation, for 8-inch pipe at 450F, per lineal foot.

A historical note is of interest here. Since the cost of installing insulation is a major part of economic evaluation, the change in dimensional standards of pipe insulations made about 1944, was an important one. The producers of the newly introduced calcium silicates rejected the old "standard" and "double standard" thickness system which was, in effect, a 1-inch and 2-inch thickness. Chemical and refining plants characteristically make many changes in piping, and when corrosion is a concern, pipes are removed or opened at key locations for inspection. Even with reasonable care, much of the thermal insulation becomes damaged so that it cannot be reused. Therefore, a large supply of replacement insulation must be maintained, both single and double layer, for all sizes of pipe. The two calcium silicate producers, quite independently, decided upon a system that used, as the base for standard thicknesses, the outside diameter of the insulation rather than the outside diameter of the pipe to be insulated and an arbitrary thickness. This provided a system of dimensions in which both the inside and the outside diameters corresponded to standard iron pipe sizes, with clearance allowances. Consequently, one layer of insulation would fit over another layer so that multiple layers could be nested to provide the total thickness desired. In the nesting system, the thickness of the insulation is dictated by the differences in standard pipe sizes, so that published thicknesses are nominal, and the actual thicknesses vary slightly from less to greater than nominal. Thermal performances, however, are based on actual thicknesses. Subsequently, the engineers of these manufacturers learned that the nesting system for pipe insulation had been proposed earlier by Raymond Thomas and his assistant William C. Turner of Union Carbide Company, South Charleston, West Virginia. Although the idea of nesting had been developed independently in three organizations, it was agreed that credit for the innovation should be given to the engineers of Union Carbide Corporation at South Charleston. By using nesting dimensions, UCC was able to reduce the number of sizes in stock from almost 700 to less than 300. Standardization of nesting led to prefabrication dimensions for rigid and semirigid material, which the industry adopted as Recommended Practice for Prefabrication and Field Fabrication of Thermal Insulating Fitting Covers for NPS Piping, Vessel, Lagging, and Dished Head Segments, ASTM C-450 (10). Nesting sizes that led to prefabrication on the job site as well as in the factory was a major factor in restraining increases in costs of "insulation installed" in the past ten years.

In housing, the increasing interest in electric heating raises questions as to the amount of insulation required. A procedure for evaluating the economics of insulation for different means of heating, and especially, electric heating, was presented by R. L. Boyd and A. W. Johnson in

1962 (6). Others have written on this subject, and data are available from electric power companies.

From practical considerations, economic analyses for all houses is unrealistic. Histories of regional weather characteristics have been compiled by the U. S. Weather Bureau. Maps showing typical heating and cooling loads are available from house insulation producers, and also are shown in the ASHRAE "Handbook of Fundamentals" (3). From these average heating and cooling loads and other factors such as number and type of windows and doors, shading, and orientation, building insulation may be planned for constructions classed as minimum, average, or superior thermally. For buildings, computer programs have been developed which enable virtually all factors to be included in heating and cooling designs.

A recent book (7), on thermal insulations and systems is the most inclusive of any published. Not only are fundamental principles discussed, with consideration of the simpler aspects of computations of heat flow, and tabulations presented from sources similar to those used here, but illustrations are given of practices in application of many materials that constitute thermal insulation systems, especially from the viewpoint of a large industrial user whose operations range from cryogenic to refractory with special emphasis on fire resistance.

III. Properties of Insulation Materials

An aim of this presentation is to make clear that insulating materials must be evaluated for properties other than the thermal resistance, R, or conductivity factor, k. Many properties, of greater or less importance, should be considered, not only for the insulations themselves, but for all associated materials. General properties are treated first in the discussion that follows; they should usually be reviewed in the order shown. Then an alphabetically arranged checklist is given, covering other properties less generally applicable to all materials. Methods of testing for these properties are described wherever possible.

A. GENERAL PROPERTIES

1. Handleability

Insulation must withstand the forces on it induced in shipping, transporting to the point of application, and attaching. In industrial installations like that shown in Fig. 1, not only the compression from the banding, but also the cutting shear of the wires must be considered. The capability of the insulation to withstand forces from the point of finishing, even within the manufacturing plant, to final attachment, is called *handleability*. So many different kinds of forces are applied in handling that no single test procedure is suitable. Some of the tests that contribute to a partial evaluation of handleability will be discussed in this section.

Tests for properties such as compressive strength, flexural strength, abrasion, and impact resistance, listed in ASTM (1) give partial evaluations, but no combination of such tests is applicable to all insulations. Moreover, low values in some respect for one type of insulation may not be as significant as for another, because high values in other properties may be compensatory.

In some respects, handleability of insulation material overshadows other properties. If the material cannot be applied with reasonable care, and if it does not remain in place to perform as expected, low k factor alone is of little practical value.

From the time insulation is made, it is subject to many kinds of abuse not only by the applicators but by others to whom this "outside" material is a nuisance. Attempts to produce lighter material, if and when this results in greater thermal resistance, often leads to a lowering of the strength factors which affect handleability. All-inclusive standards are difficult to establish. Is it reasonable to expect that accidental dropping should not cause breakage, or how far is a reasonable drop? Other characteristics may affect the evaluation. Is flexible material that is not damaged by dropping, but is so unwieldy that two men are required to handle it, advantageous? Material that has a tendency to break may also be less affected by wetting or by other system conditions.

Most insulations are covered after they are attached, either to protect them or for good appearance. Factory application of adhered jackets would often improve resistance to handling forces and mild abuse, and reduce the cost for applying the jacket as an entirely separate material, provided that relatively long lines without branches are to be insulated. The integral application of jackets or vapor barriers in sheet form to the insulation may also prevent adverse effects. However, consideration of the almost unavoidable small openings at the edges for attachment may suggest that separate application of the vapor resistance sheet is preferable.

Under the section on strengths, reference will be made to some of the aims of tests to indicate good handleability.

2. Compatibility

Most insulations are compatible with iron and steel, but if they are to be used with other surfaces, an evaluation should be made, especially for stainless steel, on which stress corrosion may occur. An example occurred in a new plant with stainless steel vessels and pipes which was completed but never operated; after two years of idleness, it was found that stress corrosion had rendered the metal unusable.

Compatibility of a covering over insulation must also be considered. An aluminum-sheet cover over an insulation that is alkaline may deteriorate to failure in a few months, since moisture is almost always available. A "breaker" sheet to separate the aluminum from the insulation, often adhered to the metal sheet, is necessary.

3. Maximum Service Temperature

All insulations are limited to a temperature range that permits them to function acceptably. No simple method evaluates temperature limits, called working or service temperatures, because shrinkage of the material under a condition that simulates actual service is in itself a behavior for which acceptable values must be defined. Two factors must be considered. One is that metals onto which the insulation is applied expand, whereas the insulation may shrink, so that in service there must be a means for dealing with the joints that open between pieces of insulation. On hot surfaces, there would be some opening even if the insulation did not shrink at all. The second is that only that portion of the insulation in contact with the hot surface, or facing it, is subject to the operating temperature; within insulation, temperature decreases rapidly with distance from the hot face. Thus, placing the insulation for a test in an oven that would heat the whole mass would not be realistic. When more than one layer of insulation is necessary to provide the desired thermal resistance, the temperatures within the first layer are increased greatly over single layer, so that shrinkage in the inside layer occurs to a greater depth. In some types of materials suited for temperatures above 600F, shrinkage is manifested by small cracks running randomly over the surface, but since these cracks are small and shallow they do not affect the thermal resistance of the material substantially and may actually increase its resistance if some water of crystallization is driven off. Obviously, there is no precise method for determining maximum service temperature; two heating procedures in laboratories are (1) soaking heat in an oven since it develops comparative data, and (2) the application of the insulation to a heated surface, as described later in discussion of measurement of shrinkage.

Temperature limits imply working temperature limitations that usually, but not exclusively, apply to hot services. Although the term is indefinite, a working limit generally means that temperature above (or below) which the material no longer provides the performance expected in normal use. Normal uses imply moderate loads for the type of material, limited vibrations, and thermal stresses in a nondeteriorating atmosphere.

Measurements of dimensions, and observations of surface appearance for checking, cracking, and warping are made after exposure to incremental temperatures continued until a definitely undesirable change has occurred. If deemed necessary, strength tests may be conducted on materials before and after exposure to working temperatures to determine if the degree of deterioration is objectionable.

After given duration of exposure to a test condition, resulting changes in the material are inconclusive by themselves. For materials with years of performance history, correlations between test observations and field behavior are significant. For new materials, some experience in service must be gained before the significance of changes in appearance or strengths during short-time test conditions can be evaluated as indicative of service performance. Factors of performance that are related to temperature limitations with some reliability for some types of materials are usually not directly applicable to other types. Therefore, several factors must be considered for their significance in deciding whether or not an insulation will perform adequately in a specific service.

4. Fire Hazard

No material is fireproof unless the temperature of the fire that can be resisted continually can be defined. To minimize the ravages of fire, materials and constructions that provide *fire resistance* are used. For thermal insulation systems, the aim is to minimize the effects of combustible materials by awareness of their characteristics as hazards. Smoke is often more hazardous than temperature, especially to persons seeking exits.

Inorganic materials do not contribute to fire hazards in the usual sense. They are considered to be incombustible—a somewhat undefinable term—and at least do not usually contribute to the spread of fire. However, in process plants carrying some types of flammable liquids, slow leaks of fluid, as from a valve packing, may wet the insulation when the temperatures are high enough to cause the fluid to ignite without an applied flame, a process sometimes called autoignition. Some entirely inorganic insulations permit autoignition at appreciably lower temperatures than others.

Inorganic fibers bonded with organic adhesives may be susceptible to a form of combustion called smoldering. Although open flaming seldom occurs, when the insulation is enclosed between incombustible walls, the smoldering is difficult to extinguish. When excess amounts of organic binders are used in some fibrous blanket-type insulations, especially if they are enclosed with sheet covers, the form of combustion called smoldering may occur. In order to evaluate this tendency, a test procedure (1) measures temperature rise within the blanket while it is in contact with a metal plate heated to the proposed maximum operating temperature of the blanket.

Organic coatings on insulation, especially pipe insulation, may be a hazard for spread of fire unless they contain fire-retarding components. Before fire-retarding components were developed, it became clear that there was need for coatings that were not such a hazard that a fire could run along the pipe faster than a man could run after it.

Paints and other coatings used as part of insulation systems may be evaluated for fire resistance by procedures described in several parts of the ASTM Standards.

Nothing is fireproof. Materials resist fires of specific intensities for varying periods. Thermal insulation systems provide fire resistances for expected types of fires and for expected durations. Two of the principal evaluations needed are, first, the degree of protection afforded by the insulation system to the equipment and its contents, and, second, the potential contribution of the system to the intensity and duration of combustion. Both the insulation and its attachments and coverings must be considered. Combustible, as a description, has not been defined in terms applicable to all constructions, of which thermal insulation systems are one. Therefore, no single procedure for evaluating fire behavior in all kinds of fires is available. However, some methods to evaluate surface flammability and spreading tendency, with indications of fuel contributed, are available (1), and other methods are in use by organizations such as Underwriters' Laboratories, Inc., Factory Mutual Laboratories, National Bureau of Standards, National Research Council of Canada, some research institutes, and several insulation producers.

B. Specific Properties and Test Methods

A number of properties are of concern to some users of thermal insulation, and these will be discussed in alphabetical order. All properties that may be encountered are not included here. Many of the properties are not independent; that is, for a class of material, when one property such as density is fixed, other properties are inherent, and cannot be selected for specification per se. Therefore, designers select insulations that meet their primary needs and then evaluate unusual operating conditions to determine need for other properties. In general, regularly produced insulations of known properties produce the lowest-cost systems, which may be called "usual." When special needs are encountered, the design of the system as a whole is altered in some details to meet specific requirements that will minimize operational problems, even though initial costs may be increased.

In evaluation of the general types of thermal insulating materials, testing for some physical properties is appropriate for all materials, but many properties need not be considered for most services, especially for materials that have an adequate experience record. Test methods given in the listing are usually those found in the 33-volume "Book of ASTM Standards," most of them in Part 14 (*1*).

Absorptivity, fluid

The amount of liquid (or gas) that will be absorbed during immersion is usually expressed as percent by weight. Insulations should not be allowed to become wet but they do, sometimes while still in the shipping package. Absorption, as a property, is usually determined by testing with water. In service, other liquids, especially the flammable ones, may be absorbed and contribute to fire hazard, although not necessarily to an objectionable degree. If a combustible liquid is absorbed into insulation and fire occurs, the liquid within the insulation may not contribute to the fire as much as the same amount of free liquid; time is required for the liquid absorbed within the insulation to come to the surface to burn.

Adhesion

The ability to develop adequate bonds, either between cements and insulation or between cements and metal surfaces, is determined from tension tests and is expressed as a load per unit area.

Adsorptivity, vapor

The amount of vapor that penetrates open porous materials and adheres to the internal surface area is expressed in percent by weight. Some materials have such extensive internal surfaces that at normal temperatures with high relative humidities, such as sometimes occur in storage in hot damp regions, adsorption may be as high as 10% by weight, but even for these materials, the usual adsorption is less than 5%, which is often negligible. Such measurable increases in weight from moisture in air may be surprising, but it should be noted that the internal surface area may be 200 acres within one cubic foot of material. A secondary effect may occur if insulation with adsorbed moisture is cooled, as during the night; some of the vapor then condenses into liquid. If this situation is repeated without alternation of a corresponding drying condition, the material may become wet even though liquid water is not present externally. One way in which moisture may manifest itself is by the formation of fungus mold on the adhered canvas jacket, due to the organic adhesive used to hold the jacket to the insulation. It is a good practice to use an antifungicide in the adhesive.

Alkalinity or pH

Few materials used as insulations are acid, so the concern is alkalinity, as pH. When insulations become wet, not only accidentally, but if wet cements are used in conjunction with them during installation, the alkalinity may cause attack on associated metals, especially aluminum. Alkalinity is not usually a basis for rejection of a material, but rather indicates a need to consider its behavior in the insulation system.

Capillarity

Usually, capillarity is not of major concern, largely because the distance liquids are caused to move through the material by capillary action alone is small, as two or three feet. However, if access to the insulation by undesired liquids may occur occasionally, the capillarity may need evaluation. Capillarity is considered in conjunction with absorption and adsorption.

Chemical Resistance

Acids. When acids, either as vapor or as liquid, may come into contact with the insulation, an evaluation of rate of reaction may be desired. No specific test methods nor time of duration have been established.

Alkalis. Some insulations have resistance to serious attack to acids, whereas others resist some alkalis.

Solvents. Contact with solvents, either as vapor or liquid, may dissolve or weaken bonding agents, especially some organic adhesives.

The problem of chemical reactions applies not only to insulations, but to the associated materials, such as attachments, coatings and coverings.

Combustibility

Combustible material has not been defined precisely, but the burning tendency of any part of the insulation system must be evaluated for fire hazard as used. Some fire exposure and fire hazard tests have been developed (1), but all situations which may arise in practice have not been evaluated. To do so would be expensive.

Corrosion

Insulations that are slightly alkaline generally do not promote corrosion on iron and steel. However, trace quantities of soluble chlorides are sufficient to cause serious stress corrosion of stainless steel, even to the point of failure in a short time. Usual chemical analyses are not adequate to detect these small quantities of chloride and resort must be made to a demonstration-type stressed-specimen developed and improved to shorten the test period by the DuPont Corporation. An example of the small amount of chloride that can promote stress corrosion is the salt in perspiration of a workman's hand, which can under particularly reactive conditions, be sufficient to initiate deterioration.

Potential corrosion of any metals that come into contact with insulation, with associated materials, or with other metals that are part of the insulation system must be recognized; e.g., stainless steel screws holding aluminum bands over insulation may react corrosively if wetted frequently, whereas iron may not, even though unsightly rusting occurs.

Density

Apparent density as such, usually expressed in pounds per cubic foot, pcf, is not usually a significant property of insulation. Figure 2 shows that, roughly, materials with lower density provide greater thermal resistance. However, this relationship must not be taken literally for insulations within some classes of material. For example, fibrous mineral wools, which may be made from slag, rock, or glass, increase their thermal resistance as density is decreased until an optimum point is reached, after which further decreases in density reduce the thermal resistance; in other words, the lowest density for a specific fiber diameter and composition does not provide the so-called best insulative property (Fig. 8). The reason is found in part in the convection within the fibrous structure, because the fewer the fibers the greater the heat transfer within the interfiber spaces. If the space between fibers is evacuated to a very low

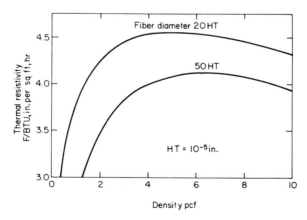

FIG. 8. Effect of density of insulation on the resistivity of two fibrous insulations with different average fiber diameters.

pressure, as 10^{-5} Torr or lower, then only negligible convection can occur, but this low vacuum is not practicable for most uses. Fiber diameter and composition and fiber-binding adhesives are factors affecting thermal performance. Heat flow tests of the insulation to be used are necessary for reliable evaluations.

Nevertheless, density is a very useful property for quality control by a manufacturer, and is simple enough for purchasers to determine as a rough measure of property acceptance. For a specific type of insulation from a specific manufacturer, job-checked density is an indication that the material has been produced as represented. If density is lower than expected, for most materials it may be presumed that the physical strengths will also be lower so that handleability will be less than desired.

Dimensional Stability

The performance of thermal insulation and the associated materials depends to a great extent upon the stability of the system during the expected life. As mentioned previously, on hot surfaces most insulations shrink while the metal being insulated expands, so that openings develop between adjacent pieces of insulation, both longitudinally and transversely. The extent of such openings that can be tolerated must be determined by the operating departments. The concern over joints on hot surfaces is not heat loss, because that is small, but rather deterioration of coverings and the fire hazard of hot surfaces exposed to fluids with a low ignition point.

Insulation without any shrinkage, or especially with some expansion would be desirable, but there are few such materials. In general, acceptable shrinkage is about one percent of length, but this is still a substantial amount, almost ⅜ inch in 36 inches, the standard length of most industrial insulations. The types of tests which yield meaningful values depend upon the principal form of the material. One kind of shrinkage characteristic is evaluated by heating in an oven, called soaking heat, and another by application of one surface of the insulation to a surface heated to the temperature expected in service. Although oven heating tests do not expose the material in a manner related directly to service, the test values are indicative of temperature behavior to be expected. A concern is the question of test exposure time which may bear a reliable relation to field performance.

Stability against temperature is an obvious need; stability against moisture or even wetting is of equal or of even greater importance. It is hoped that wetting will not occur; if it does, and later the insulation freezes, disruptive forces within the insulation are invoked. The expansive force of ice formation is well known.

For some types of materials, such as organic fibers and foams, stability in high relative humidities, especially when combined with warm temperatures, should be given special consideration.

Dimensional Tolerance

Dimensional tolerances are the acceptable deviations from standard dimensions. So-called rigid and semirigid insulations are applied as separate pieces with joints between them, and heat transfer through the joints by radiation and convection may be objectionable. Undue dimensional tolerance in a material tends to increase the size of the joints. The total heat loss through joints is usually so small that it is negligible, but radiant heat may raise the temperature of the covering or the coating to a hazardous point. Similarly, when an insulation system is installed, it should have a neat workmanlike appearance. If adjacent pieces of insulation are of different thicknesses, an irregular outer surface is inevitable. Years ago, it was common practice to fill the joints and to level the outer surfaces with wet plastic cements, but this increased the installation cost appreciably and introduced water into the insulation. If this work was done shortly before the system was put into operation, the wet insulation would not have the thermal properties expected until it became dry, and the moisture being dried might contribute to corrosion of metals upon which it condensed. Manufacturers are finishing their products with greater attention to dimensional tolerances than in the past, including warping. Measurement for different types of insulations are in the voluntary standards of ASTM (1).

Expansion Coefficient. See Linear Coefficient of Expansion-Contraction
Hygroscopicity. See Adsorptivity.
Indentation Resistance. See Strength.

Linear Coefficient of Expansion-Contraction

Most insulating materials represent a compromise between desirable and less desirable physical properties. Thus, characteristic changes in length due to temperature, especially for services above ambient, even though opposite in direction to that desired, are tolerated if the induced changes result only in cracks or in openings of less than about ⅛ inch between standard pieces of insulation. Two aspects of system behavior tend to minimize the need for critical consideration of thermal coefficients of length change of material. First, for many hot services, only a small portion of the material becomes heated to the service temperature, so that the linear coefficient is not applicable literally. Second, the thermal coefficient of the metal surfaces to which or near which insulating materials are applied are much greater than the linear coefficients of the insulation. Consequently, installation techniques to compensate for the length changes of the metal being insulated also compensate for the changes in insulation. Some insulations in rigid form and some of fibrous form have little or no shrinkage characteristics below their recommended operating temperatures, but the dimensional changes of the metals being insulated usually govern the joint or crack opening problems.

Moisture Resistance

Thermal resistances and k factors as published apply to dry materials. When insulations become wet internally, either accidentally or by migration and condensation of moisture, their thermal performance is often reduced. This relationship of moisture increase and decrease in performance has sometimes been misinterpreted to mean that any moisture at all induces major increases in heat flow. In fact, many insulations that at times may contain small amounts of moisture, say 5% by weight, do not change in heat transfer rate enough to warrant practical objections. However, this comment should not be considered an excuse for not installing and maintaining virtually dry insulations. Moisture that enters during nonoperating periods is usually driven out when surfaces return to heated operations. However, if moisture is allowed to migrate into cold insulations, the decreasing thermal resistance induces even greater ingress of moisture so that degradation in performance and eventual failure occurs. Therefore, operations at temperatures below ambient require most careful design, installation, and maintenance to avoid moisture-related deterioration.

Since these same laws of moisture flow within materials apply to test specimens, it is virtually impracticable to measure reliably the rates of heat flow through wet specimens.

As mentioned elsewhere for emphasis, moisture resistance of the insulation system must be sufficient to keep the insulation material dry and to avoid associated corrosion. Sometimes the concern is not only with moisture within the insulation, but also with moisture in the migration paths of the joints between the insulation pieces or between the insulation and the envelope.

Procedures measuring moisture migration through materials in either sheet or thicker form have been developed (1) for specific test conditions. A danger lies in the supposition that the behavior seen in tests will hold for a different exposure condition in the field, or that the seemingly low rate of moisture flow which the test results express per hour or per day may not become objectionable over long operating periods under conditions that induce a continual inflow of moisture.

Odor

Odors associated with thermal insulations are usually of little concern in industrial systems, but they need consideration for house insulations and especially for food storage systems.

Permeance. See Water Vapor Resistance.

Radiation Resistance

In atomic energy applications to minimize heat transfer, insulations that are inorganic and do not contain reactor poisons have been found not to deteriorate objectionally in strength, and their thermal resistance has been adequate. Occasionally, the use of all-metal reflective insulations in locations in which high surface temperatures are not objectionable may provide ease of removal for inspections with high reuse capabilities.

Specific Gravity and Specific Heat

Since specific heat, in conjunction with either specific gravity or density, provides a calculated value for thermal diffusivity, a necessary factor in transient heat exposures, evaluations of these properties are often published by manufacturers. Several methods for evaluation at ordinary temperatures are published. Special techniques are used for obtaining values in the cryogenic or the very high temperature ranges.

Shrinkage. See Linear Coefficient and Temperature Limits.

Strength

Values for strength, as indicated by different types of loading, aid in evaluating handleability and physical performance, and some minimal values are defined in part in specifications (1), which also refer to the testing procedures to be used. Widely differing materials may be suited for the same operating condition, and if their handleabilities are acceptable, it may be unrealistic to compare them literally on the basis of strength unless a particular mode of strength is part of performance needs. However, published strength values have a use often overlooked; the particular strengths may not be essential in a specific application, but the values are useful for evaluation of acceptance quality, to assure that the material at hand was made to the quality level represented. In the following comments on strengths, differentiation is necessary between values for design loads and for indicators of quality assurance.

Compression. Most insulations are not brittle and deform slightly under load, so the purchaser must specify the allowable percent compression, which will define how the strength is to be evaluated. In most cases, the insulation need support only its own weight and the coverings and attachments. However, if the insulation is expected to support people walking, or equipment on a permanent or a temporary basis, the compressive strength with a factor of safety must be adequate for the loads, not only for dry insulation, but also for potentially wet insulation.

Flexural strength, modulus of rupture, breaking load. These terms are closely related and indicate values from the application of loads that bend the material as a beam. Flexural strength is one of the indicators of handleability, and is a simpler test than compression for quality acceptance. For most insulations that are stronger in compression than in tension, the flexural loading is a form of tension evaluation. Insulations that absorb appreciable amounts of water are often reduced in flexure resistance when wet, so that their ratio of wet to dry strength may be an indicator of proper manufacture and be thereby suitable for quality assurance.

Tension. Direct measurement of tensile resistance of insulation is seldom of concern. Flexure loading is a form of tensile loading that aids in evaluating handleability. However, tensile strength of auxiliary materials may be of interest for quality assurance.

Shear, punching shear, and indentation resistance. Many loads applied to insulations are compressive. If the area of application is small, the stresses within the insulation are primarily shear at the edges of the load, and if the loads are applied on areas as narrow as the thickness, the forces may be punching shear or indentation. Walking or placing ladders or scaffolding on or against covered insulations may produce enough shear or indentation to harm, not so much the insulation itself, but the covering. Even tears too small to see will permit the ingress of water or vapor. This situation may not be serious on hot operations, but it will be of concern on cold surfaces, the colder the more serious. At low cryogenic temperature, moisture ingress could induce failure of the insulation system.

Impact, drop test. Since other test values alone are not entirely satisfactory to define handleability, another procedure subjects blocks or segments of pipe insulation to impact loading from a swinging pendulum hammer, or by dropping the specimen in a guide frame to determine the height of drop to cause fracture. Suddenly applied loads induce different kinds of stress distributions within materials than do slowly applied loads, called static loads. Therefore, dynamic loading has use on new materials for which field history is lacking.

Thermal Shock Resistance

Some insulations that have many desirable properties are unable to resist the stresses induced by sudden changes in temperature, called thermal shock. When one surface is heated or cooled too rapidly, the induced length changes create stresses beyond the strength of the surface so that cracking occurs. If the shock is repeated, progressive cracking will lead to failure. Most resilient materials resist thermal shock, although distress may occur if the shock is accompanied by the adverse effects of migrating moisture. No general test is available.

Tumbling Resistance

A test for tumbling resistance is an indication of one aspect of handleability for some rigid insulations that can be sawed into small blocks. When a number are placed into a wood box and rotated so that the blocks tumble over each other, the induced weight losses by impact and abrasion on the edges of the blocks indicate the resistance of the material to the forces of shipping and application.

Vermin Resistance

Insulations are not resistant to boring of vermin seeking to pass through walls and other insulated structural components, but most of them do not provide food for vermin to encourage entrance. However, some circumstances may lead to nesting. Antivermin treatment of the insulation is usually not necessary, but may be desirable locally.

Vibration Resistance

Industrial insulations may be subjected to vibrations when they are associated with machinery, or when they provide thermal resistance on equipment for which vibration is a function. Tests for vibration resistance were conducted at one time by U. S. Navy laboratories, but they have not been used recently there or elsewhere, largely because equipments are now designed to avoid vibration. When vibrations are encountered, attention to method of attaching the insulation so that its motions are integral with the attached surface, and so that looseness of fit does not induce rattling, is usually sufficient for acceptable performance of materials that possess some resilience.

Water Vapor Transmission: Permeance

Water vapor is usually not of concern in insulation systems that operate continually at temperatures above ambient. However, during nonoperating periods, migration and condensation of moisture may lead to deterioration of the material or may lead to corrosion of associated metals. Since many insulations of light density are also open pored, their resistance to vapor flow is negligible. Some closed-cell materials do have high resistance to moisture transmission, but virtually none are absolute barriers when joints and unavoidable openings are considered. Consequently, most tests for vapor transmission are directed to the coatings and coverings to be used with the insulation, or to laminates of vapor-resistant material with the insulation. Permeability is a property of material as determined under standard conditions (1) for unit thickness; permeance is the moisture transmission rate of the material as used and is the unit usually used for evaluation of materials. A low permeance material is a moisture migration retarder. In Fig. 11, the low permeance sheet being spread on the ground retards appreciably the migration of moisture from the ground into the house above.

The function of materials to restrict flow of moisture must be evaluated for specific constructions, especially when the vapor pressure differential, which is the natural force that drives the moisture, may reverse direction. Field vapor transmission rates into or out of systems that depend upon weather usually change appreciably and sometimes reverse, so that periods of drying tendency exist. In thermal insulation systems such as the walls of houses, which cannot be absolute barriers, the moisture that can enter from the warm side should meet so little resistance that it can continue to flow entirely through the wall to outdoors, as a flow-through principle. Therefore, the construction must provide no more moisture resistance within the wall than is provided at or near the warm side. To assure that moisture that reaches the outer surface of the wall will actually flow to outdoors, the resistance of the outer elements must be appreciably less than anywhere within the wall. The permeance of the outer surfacing should be appreciably greater, on the order of five times or more, than the permeance at the warm interior.

For cold-storage and freezer systems, and certainly in the cryogenic range, the vapor barriers must be reversed in location, that is, placed on the warm side or outdoors. In these cases, not only low permeance constructions but literally absolute barriers are essential. These cold constructions require not only knowledgeable designers, but experienced constructors and installation personnel.

IV. Component Materials of Insulation Systems

So many materials with such widely varying characteristics are available that an all-inclusive list is beyond the scope of this discussion. All of the presently used materials, and new ones yet undeveloped, perform under the same physical laws of heat, moisture, and stress. The following sections describe many of the kinds of service encountered; some materials are suited for a particular service and some are not. More than one type of material may be suitable for the same operating conditions, but no two materials have exactly the same properties even though their thermal resistances are the same at some temperature. In practice, it is usually uneconomical to select for one insulation system, several materials, all of which are competitive for the same service, solely because each one gives slightly better performance in some particular respect. Compromises on noncritical properties so that the least number of different materials are used would reduce installed costs.

To reduce the number of different materials used, compromises may be made between thermal resistance and thickness, handleability, and resistance to breakage or to accidental exposure to wetting, fumes, spillage, or abuse. Moreover, use of different insulations in the same work area introduces confusion, so that materials suited for one surface are placed on others where they do not belong.

Only general indications of properties are presented here as illustrations of service suitability. For information on properties of a particular brand, the literature of the producer should be obtained, since published general properties (3) are inadequate for designing to fit unusual performance requirements. Moreover, a currently produced material may be an improved version of an older material of the same name.

A. SCHEMES OF CLASSIFICATION

So that designers can determine which insulations should be considered for a new process or new construction, classification of materials in general terms is needed. For this introduction to thermal insulations, some of the more common classification schemes are described. Then a descriptive list is arranged alphabetically under classification by shape and form of the material as it affects handling and installing. This arrangement emphasizes that the choice of materials is to be made by knowledgeable purchaser-designers to construct well-engineered systems.

1. Service Temperature

Insulating materials are often classified according to temperature limits. Refractories are obviously thermal insulators; their range of service is considered as upward of about 1800F (1000C).

The cryogenic range was at one time considered to be below the liquefying temperatures of the gases of the air. General references to cryogenic insulations now imply service below about −100 to −150F (−75 to −100C).

The term "thermal insulation" is usually used to imply temperature ranges between cryogenic and refractory. It is obvious that these temperature classifications overlap widely, but they serve to indicate the design factors encountered. In industrial operations, all three temperature ranges are common.

Classification of thermal insulation for smaller ranges of service temperatures is normally made by the engineering staffs of industrial users who have need for several types of materials.

Unfortunately, the possibility of exposure to fire temperatures, many of which are higher than service temperatures, has not received the attention which good practice would demand; in choice of materials, compromise between lesser thermal service performance and capability to resist some types of fires would save major parts of constructions from substantial losses.

2. Construction Application

Thermal insulations are roughly classified according to probable sources of materials, because no manufacturer produces all of the types of insulation on the market. Sales distribution organizations are provided by manufacturers for local needs ranging from small quantities of materials to occasional truckload and carload quantities. Multicarload quantities are shipped directly from factories to jobsite to save large storage and rehandling expense. Since most needs for insulations are not so restrictive that only one material for a particular service will be acceptable, ready availability of materials at competitive cost enables consumers, or their designers, to use some freedom of choice. From a construction standpoint, thermal insulations are classified broadly as industrial, commercial, and residential. Generally, the materials for commercial and residential use are the same, but the quantities per purchase are relatively small for residential use and are available over the counter, although the quantities for large tracts of residences are multi-truckload. Even for industrial consumers, local availability for immediate delivery minimizes the need for consumer stockroom storage of insulations and accessories for repairs and minor changes in operations. Ready availability is expected for commercial constructions, stores, warehouses, shopping centers, office buildings, etc.

3. Internal Structure

Insulation materials are characterized by variety in internal structures.

a. Porous Solids

Although porous solids have the appearance of solid material, their internal structure consists of particles interspersed with minute spaces, which may or may not be interconnected. Most materials of this class are inorganic, and are suited well for temperatures from moderate to high, 250F (120C) to 2000F (1100C). According to their physical properties, porous solids are classified roughly as rigid or semirigid. Cork is a rigid organic material, usable only for temperatures below 200F (90C). Heavy density organic foams, although basically porous solids, are usually considered a class by themselves on the basis of composition.

b. Fibrous

Fibrous insulations, whether organic or inorganic or combinations, depend primarily upon minimizing internal convections and radiations by creating small interstices with interposed fibers of such small diameters that they are described by industry habit in HT (0.00001) hundred-thousandths of an inch, or 0.000254 mm, and are oriented insofar as practicable so that conductances along the fibers are primarily perpendicular to the direction of heat flow to be resisted. Fibers may be coated with adhesives so that they adhere when they touch and set so that the mass will be more readily handled, or they may be left as uncoated fibers which hold together slightly by intertwining. Some fibrous insulations are pelletized so that they can be poured into cavities, as in walls or between joists.

Inorganic fibrous. The more common insulations are made by fiberizing from the molten state rock, slag, or glass, and fabricating the fibers with or without binders (adhesives) into various forms from low-density blankets (0.2 pcf) to heavy boards (16 pcf) which are quite rigid. Low-density materials are presumed by many of the uninformed to be "better" insulations. Although decreasing density from 16 pcf usually does increase thermal resistance, an optimum density for a particular fiber diameter and method of forming is reached, and then further decrease in density reduces thermal efficiency. The optimum density is a characteristic for each material, as exemplified in Fig. 8.

Asbestos, a peculiarly fiberizable rock has been known for centuries. Since some of its properties have been misunderstood, it will be described in detail with some history in a later section.

Other somewhat exotic fibers with high temperature resistance are available for specific services but are limited in production volume by cost.

Organic fibrous. Natural or synthetic fibers such as animal hair, sugar cane (bagasse), ramie, wood, cotton, animal wool, and several types of synthetic thread are formed into blankets or batts, or left in a bulk form packaged suitably for shipping and handling.

c. Foam or Cellular

As the term implies, the basic material, whether organic or inorganic, is expanded to a much lower apparent density than its solid state by conversion into myriads of tiny bubbles or cells so that the walls between cells are very thin. The cell walls may or may not be continuous so that the interstices may or may not be interconnected; interconnected cells provide a lowered resistance to vapor migration and are less desirable when moisture ingress is to be avoided, but they provide desirable acoustic properties. However, even when cell walls are continuous, they may be so thin that some gases may pass through them.

Inorganic Foam. The principal inorganic material of this class is foamed glass, made from a mixture of foaming agents and glass granules heated to incipient fusion so that the mass rises in a pan like a loaf of bread. After baking, cooling, and stripping from the mold, fabrication into desired shapes is accomplished by sawing or grinding. More intricate shapes may be formed by cutting and cementing pieces into larger units such as equipment or valve covers. For aero-space use, porous foams of silicon carbide, alumina, zirconia, and silica when suited to the 3000 to 4000F range could be classed as porous solids, but, as stated previously, foams have a separate designation.

Organic Foam. Plastics that react quickly during expansion of gases through the mass, and then solidify around the gas bubbles are known as plastic foams. Like inorganic foams, plastic foams may or may not be structured with connected cells, so that for a particular product the rate of vapor transfer, especially water vapor, needs evaluation for some applications. Plastic foams are available in a range of densities. A compromise of physical properties is found in a density of about 2 pcf and is quite commonly used in the building industry.

d. Granular

Pellets of various materials, such as vesicular or exfoliating rock, or synthetic ceramics, expanded vermiculite (mica), expanded perlite, pelletized foams, pelletized fibers, etc. are suitable as loose-fill insulations. Some may be coated with organic or inorganic adhesives (binders) and formed into units for ease of handling, or they may be formed in place in structures moving the top open but the sides and bottom restrained by structural supports.

e. Reflective

Solid metal rolled into very thin sheets, called foils when less than 0.005 inch (0.127 mm) thick, reflect heat because of their characteristic property of reflectance, or restrain heat from escaping due to their characteristic property of emittance. To be fully effective, the surfaces must be highly polished. The terms reflec*tivity* and emmis*sivity* are reserved for the ultimate values attainable for a material as an inherent property. The terms reflec*tance* and emit*tance* describe the performance of the materials in the condition as used. Commercially polished surfaces are not necessarily the ultimate, and in use, dust and other films reduce the performance to lower than ultimate values. To be fully effective, reflectance surfaces require free space, which in air should be on the order of one inch for buildings. Reflective super-insulations are combinations of many reflective sheets separated slightly by relatively low conductance fibers to form 120 or more sheets per inch. However, their potential high thermal resistance is not reached until the interply space is highly evacuated, lower than the order of 10^{-5} Torr.

f. Ablative Materials

For extreme high temperatures of short duration, a special type of thermal insulation, usually referred to as ablative heat shield, is used. A characteristic of ablative material is the capacity to maintain integrity throughout all of its mass except that surface to which the sudden heat is applied, especially while this heated surface disintegrates. A common type is a plastic mass reinforced with certain types of fibers. Since the localized disintegration involves a substantial change of state from solid to vapor and there is often dehydration of particulate matter, large amounts of heat are absorbed at the disintegrating surface. Moreover, heat is absorbed by the mass not yet at disintegrating temperatures, so that for a short time the cooler side of the mass does not increase in temperature objectionably. Materials that can be sacrificed on one surface in order to preserve the integrity of the remaining mass are called ablative materials. A well-publicized ablative material is the heat shield on the bottom of the re-entry capsules of astronautic excursions. Obviously, ablative materials are consumed rapidly, so that one-time use is the usual service, although in noncritical applications re-exposure of the remaining ablative material may be practicable. The term heat shield has meanings other than as used above, such as an interposed heat barrier, as a curtain to protect personnel from high radiation rates near openings in furnaces.

4. Form and Shape

Since insulations must be installed under a wide variety of circumstances, they are commonly classified according to the forms and shapes in which materials are available. Generally, pieces of insulation are desired that can be handled easily by one workman. Occasionally, insulation sections are handled by two men when a large area of unwieldy material is being applied, or in hazardous locations, as on windy days on high scaffolds.

The standard length of insulation blocks and pipe insulation is 36 inches, although some longer lengths are produced for special services. Pipe insulation is made as complete tubes for some limited services, but the usual conventional tubular forms are called sectional for two hemi-cylinders, tri-seg for three segments, and quad-seg for four segments. For pipes and vessels of diameter larger than about 36 inches, curved segments of a single curvature are called segmental.

B. MATERIALS FOR THE INSULATION LAYER

The following list of materials, classified by form and shape, provides information on temperature limitations as well as brief descriptive comments. The approximate low-temperature specifications shown are not physically limiting, but indicate that the material is not usually used for lower temperatures. Materials are labeled as primarily inorganic (I) or primarily organic (O). Some inorganic materials are formed into units for field application by means of either organic or inorganic adhesives. In thermal insulations, such adhesives, which are not of themselves major insulations, are called binders without identification of their composition, but the effect of the binder on performance may determine the working-temperature limitations.

The list is limited to examples of materials which indicate the wide range of service conditions encountered in designs. No significance should be attached to any omissions.

1. Block, Board, and Preformed Pipe Insulation

Material	Temperature Service Range	Inorganic (I) or organic (O)
Alumina, silica with binder. Chemically inert, soft, absorbent.	70 to 2000F (20 to 1100C)	I
Asbestos fibers, diatomaceous silicas with binder. High tensile strength, good handle-ability, low thermal shrinkage, absorbent.	70 to 1400F (20 to 760C)	I
Calcium silicate (reacted) containing asbestos fibers. High compressive strength (over 100 psi), fabricates easily for fittings, resists thermal shock, absorbent.	70 to 1200F (20 to 650C)	I
Calcium silicate (reacted, special constituents). Similar to above with higher temperature limit but not as strong.	70 to 1700F (20 to 925C)	I

Material	Temperature Service Range	Inorganic (I) or organic (O)
Cork molded granules with organic binder. Good handleability, semirigid. Used for moderately cold service.	−250 to 200F (−160 to 90C)	O
Diatomaceous silica with binder. Semirigid, moderate strength, low thermal shrinkage. Often used as first layer beneath less expensive and more efficient insulations.	70 to 1900F (20 to 1040C)	I
Glass, cellular foam. Rigid, high compressive strength (150 psi), fabricates well, water vapor resistant in closed cell form (available with interconnected cells), does not resist thermal shock, resistant to many chemicals.	−450 to 800F (−270 to 425C)	I
Glass fibers with organic or inorganic binder. Semirigid to flexible, available in several densities and other forms, listed later.	−100 to 450F (−75 to 230C)	I
85% Magnesia (common name), basic magnesium carbonate. Semirigid, moderate strength. Some varieties soften when wet. Can be fabricated but somewhat dusty.	70 to 550F (20 to 290C)	I
Mineral wool fibers (rock or slag) with binder. Semirigid, moderate compressive strength, absorbent.	70 to 1700F (20 to 925C)	I
Polystyrene, expanded bead (a form of foam). Light density, usually about 2 pcf, good compressive strength per unit weight. Will not withstand fire. May need vapor barrier for joints.	−400 to 175F (−240 to 80C)	O
Polystyrene, cellular (a different form of foam). Available by cutting from larger units manufactured only by producer.	−400 to 175F (−240 to 80C)	O
Polyurethane, expanded cellular foam. Available in several densities which affect strength and rigidity. Will not withstand fire.	−20/−400 to 175/250F (−30/−240 to 80/120C)	O
Rubber foam. Flexible in light density, can be slid over piping as an unsplit tube. Will not withstand fire. Provides some resistance to vapor flow.	−400/0 to 180/220F (−240/30 to 80/105C)	O

Material	Temperature Service Range	Inorganic (I) or organic (O)
Silica, expanded granules, with binder. Good handleability, fabricates well, low thermal shrinkage, resists thermal shock, rigid, water resistant.	70 to 1600F (20 to 870C)	I
Silica, fused cellular, foam. Rigid, strong, resistant to acids and water. Recently withdrawn from market because manufacturing costs were too high for limited market into which other materials were appearing. Mentioned here as an example of performance for unit of cost, when other materials can be used.	−300 to 2000F (−180 to 1100C)	I
Silica, fused, molded. High compressive strength. Acid resistant, rigid, available in several densities.	70 to 2000F (20 to 1100C)	I
Vegetable matter board. Fibers are combined into various densities to provide handleability and strengths needed for building sheathing and similar wall boards. Some have acoustical properties of use in ceilings and within walls. Fibers may be cane, wood, or reused paper; each has characteristic properties.	−60 to 150/200F (−50 to 65/95C)	O

2. Reflective Metals

Material	Service Range	Inorganic (I) or organic (O)
Aluminum, sheets or preformed shapes. Polished on one or both sides. Usually less than 0.005 in. thick (0.12 mm) called foil. May be attached directly or shaped and fabricated. Incombustible but melts in mild fires. Thin foils may not be absolute vapor barriers due to pinholes.	−100 to 1000F (−75 to 540C)	I

Material	Service Range	Inorganic (*I*) or organic (*O*)
Aluminum adhered to mats, paper, or films. Polished side faces space. Adhered films may improve vapor resistance over foils.	−459 to 70F (−273 to 20C)	I/O
Gold, adhered to plastic films. Usually not suited to high temperatures. Radiation shield.	−459 to 180F (−273 to 80C)	I/O
Stainless steel, polished sheets or foil. May be performed or fabricated. Not as efficient as polished aluminum but has higher working temperature.	−100 to 1200F	I

3. Blankets, Batts, and Felts

Alumina-silica fibers. Flexible, available as blanket, felt, paper, cloth, and tape.	70 to 2000F (20 to 1100C)	I
Asbestos fiber with or without binder. Temperature limit depends upon variety. Available as blanket, felt, cloth, and paper.	70 to 750/1000F (20 to 400/540C)	I
Glass fibers without binder. Available as batts or blankets, or attached to open metal mesh. Omission of organic binder permits higher service temperature.	−450 to 1000F (−270 to 540C)	I
Glass fibers with organic binder. Available in several fiber diameters, densities, and facings. Flexible with some resilience. Will not withstand fire.	70 to 350/450F (20 to 175/230C)	I/O
Mineral wool (rock or slag) with or without binder. Glass wool is also mineral wool, but the rock and slag wools are usually usable at higher temperatures than glass, although there is no sharp demarcation. Characteristics of mineral source determine service temperature. Flexible blankets and felts available with or without several types of facings. Easily compressible. Adequate vapor barriers required for low temperatures.	−450 to 600/1200F (−270 to 315/650C)	I

Material	*Service Range*	*Inorganic (I) or organic (O)*
Quartz fibers. Available as felts and tapes, usually used for high temperature service.	70 to 2800F (20 to 1540C)	I
Silica fibers. Similar to glass fibers but with elements other than silica removed to improve temperature limits and resistance to chemicals. Usually used in densities less than one pcf.	−400 to 2000F (−240 to 1100C)	I

4. Sprayed, Foamed, or Foamed-in-Place Insulations

Asbestos fibers and binders. Applied by spraying by nozzle onto surface, sometimes followed by troweling. Binders react and set, and excess water dries. Resistant to fire.	70 to 700/1300F (20 to 370/700C)	I
Asbestos and selected mineral fibers with binder. Relatively dense, foamed in place. Resistant to fire.	70 to 2000F (20 to 1100C)	I
Asphalt with cork or micaceous fillers. Acts as coating with some thermal resistance, applied by brush, trowel or spray.	−20 to 200F (−30 to 95C)	O
Polyurethane foam, rigid or flexible. Foam sprayed or formed in place. Not resistant to fire.	−60 to 160F (−50 to 70C)	O
Polyvinyl acetate with granular filler. Acts as coating with some thermal resistance, some fire resistance.	−20 to 180F (−30 to 80C)	O
Rubber with cork granules. Usually in form of sheets to mold by hand.	−300 to 200F (−185 to 95C)	O
Vinyl with cork or mica. Acts as combination of coating with some thermal resistance.	−20 to 180F (−30 to 80C)	O

5. Cryogenic Evacuated Systems

Aluminum foil with glass mat or paper.	−459 to 70F (−273 to 20C)	I
Calcium silicate, cellulated.	−459 to 70F (−273 to 20C)	I

Material	Service Range	Inorganic (I) or organic (O)
Perlite. Expanded and graded perlite ore.	−400 to 350F (−240 to 175C)	I
Silica (aerogel) with copper flakes.	−459 to 70F (−273 to 20C)	I

These materials for cryogenic service must be installed in vapor- and air-tight enclosures, and should be evacuated to absolute pressure of less than 50 microns of mercury. Heat transmittance characteristics vary but are on the order of 1 to 18 $\times$ 10^{-5} Btu per hour, per square foot, per inch of thickness.

6. Exotic Insulations

Ablative. Materials which are solid until high temperatures are reached and then change state to vapor as they absorb large quantities of heat for short periods. The heat shield on the bottom of astro-capsules is an example. There can be industrial uses.

7. Loose Fill Insulations

Loose fill materials are usually in pelletized or granular form or fibrous in short lengths so that the bulk material can be poured or blown into cavities.

Material	Service Range	Inorganic (I) or organic (O)
Alumina-silica fiber. Loose fibers, 4 to 10 pcf installed.	70 to 2300F (20 to 1260C)	I
Alumina-silicate. Loose granules, 40–50 pcf installed.	70 to 2000F (20 to 1100C)	I
Asbestos fiber. Various grades and densities 20–50 pcf.	70 to 800/1200F (20 to 425/650C)	I
Calcium silicate. Usually pellets, absorbent.	−459 to 1000F (−273 to 540C)	I

Materials	Service Range	Inorganic (I) or organic (O)
Cork granules. Various grades available.	−200 to 200F (−130 to 95C)	O
Diatomite. Diatomaceous earth, fine to coarse powder, 10–30 pcf installed.	70 to 1600F (20 to 870C)	I
Diatomite, calcined. Can be mixed with cement, 25–30 pcf.	70 to 2000F (20 to 1100C)	I
Gilsonite (asphaltic). Various temperature limits, must be heat cured, usually in place.	70 to 500F (20 to 230C)	O
Glass, cellular pellets. Foam walls, sealed. 5–10 pcf.	−300 to 800F (−185 to 425C)	I
Glass fibers. Available unbonded, unlubricated or lubricated. 2–12 pcf installed.	−300 to 1000F (−185 to 540F)	I
Glass, fibrous with binder. Usually organic binder, known as pouring wool. 1–3 pcf installed.	−300 to 250F (−185 to 120C)	I
Glass, fibers of borosilicate glass. Good electrical resistance, 2–10 pcf.	70 to 1500F (20 to 820C)	I
Mineral wool, slag or rock. I. Fibers granulated or nodulated. 3–20 pcf.	−300 to 1200F (−185 to 650C)	I
Perlite expanded ore. Nodules. 3–10 pcf.	−400 to 1500F (−240 to 820C)	I
Quartz fibers. Available in several diameters. 3–12 pcf.	70 to 2500F (20 to 1370C)	I
Silica aerogel. Small hollow spheres. 4–6 pcf.	−457 to 1300F (−273 to 700C)	I
Silica fibers. Long fibers of silica. 3–10 pcf.	−300 to 2000F (−185 to 1100C)	I
Silica, fused. Granular of several sizes. 40–100 pcf.	70 to 2300F (20 to 1260C)	I
Vermiculite. Granular expanded vermiculite ore, 7–12 pcf.	−350 to 2000F (−215 to 1100C)	I

C. Supplementary Comments on Asbestos and Reflective Metals

Two insulation materials, asbestos and reflective metals, have an interesting history of development. They illustrate in different ways that the resistance of any insulation to the three modes of heat transfer—conductance, convection, and radiation—coupled with other physical properties, establishes their marketability.

1. Asbestos

Asbestos is a rock, mined underground from strata of varying thicknesses, generally not more than six inches. Many strata are less than one inch thick. Although asbestos outcrops exist, the surface material has usually been weathered and has limited use. Asbestos rock has the unusual property of being separable quite easily into hairlike fibers so fine that natural or synthetic fibers do not approach them. The diametral values tabulated below are illustrative and not limiting.

Material	Fiber diameter[a]	
	inch	mm
Human hair (fine)	0.00200	0.0508
Ramie	0.00100	0.0254
Wool	0.00080	0.0203
Cotton	0.00040	0.0101
Rayon and Nylon	0.00030	0.0076
Rock wool	0.00015	0.0038
Glass	0.00005	0.0013
Asbestos (chrysotile)	0.0000007	0.000018

[a] Some data from *Canadian Mining and Metallurgical Bulletin* (1951).

In ancient times, at least five centuries B.C., asbestos was spun into cloth that was found to resist fire, and was called amianthus (8). Plutarch, prior to 400 B.C., recorded that the vestal virgins tended perpetual lamps having wicks of a woven material that has been presumed to be largely amianthus. Amianthus from the Italian Alps was used for cremation cloth to protect the bodies on funeral pyres with "linum vivum," as the ancients called it, which probably meant everlasting linen. When no other use was found for the cloth, and the pyre fashion was discontinued, amianthus was almost forgotten. Historians find references over the years in learned societies to this magic material. Marco Polo in 1250 A.D. was shown unburnable cloth which he traced to Siberian mines.

Amianthus, rediscovered and renamed asbestos, became commercially important late in the nineteenth century because the higher temperatures being encountered in industry necessitated heat conservation.

The presence in Canada of asbestos of the white variety called chrysotile was noted in 1847 by Sir William Logan, the first director of the Geological Survey of Canada. Principal deposits were not operated until about 1880 at Thetford, Quebec, and these are still the world's largest producers. Even before Logan's time, about 1805, German geologist H. Lichtenstein discovered a lavender asbestos (crocidolite) in the Orange River valley of South Africa where it is now produced as "Cape Blue" asbestos. About 1907, a dark brown asbestos with unusually long fibers was found near Penge, South Africa. Its composition was recognized as not the same as another common mineral, anthophyllite, so it was considered to be a separate type. One mining company was named AMOSA (Asbestos Mines of South Africa) and later this name was used as a means for distinguishing the brown asbestos (color due to high iron) from the blue asbestos mined not far away. In 1918, Dr. A. L. Hall of the Geological Survey of the Union of South Africa suggested that the modified name amosite be adopted, and it is still used.

The three principal source areas for asbestos are Canada for chrysotile and Africa for Cape Blue and amosite. Several other source areas are scattered throughout the world with very little in the United States.

One of the variations in composition of different types of asbestos, all of which are complex silicates of magnesia, iron oxides, and alumina, is water of crystallization. Since water of crystallization is driven off at relatively low temperatures, none of the types of asbestos is suited to sustained exposure to more than about 1400F (760C). Chrysotile contains about 15% by weight of water of crystallization, which is roughly three times that of amosite and crocidolite. The water causes greater shrinkage and loss of strength properties in chrysotile than in the other two as temperature rises. These temperature effects on chrysotile above about 800F (425C) may be objectionable for some services. Temperature limitations indicated apply to the asbestos alone and not necessarily to products into which it has been incorporated.

Since only one surface of insulation is usually in contact or exposed to the hot surface being insulated, it is the only portion at maximum service temperature. Many inorganic synthetic and some organic-base insulations contain chrysotile, amosite, or both. Crocidolite is better suited to chemical-resisting applications; its cost is prohibitive for most thermal insulations. Asbestos fibers are incorporated in other insulation materials, not for their thermal resistance, but for the strength improvement from reinforcement by the many strands. Consequently, the length of fiber as well as its strength must be evaluated for cost per unit of strength improvement. The longer-fibered asbestos grades are much more expensive than the short.

One type of insulation produced in blocks and tubular form for pipe insulation is molded almost entirely of amosite asbestos of several length gradations, with added inorganic fillers, cemented into moderately rigid form. Another type is delivered in dry loose form and sprayed wet with a machine blower hose and nozzle. Spraying is suited especially to irregularly shaped surfaces.

As discussed elsewhere, thermal insulation systems are installed materials in various forms for handling. However, since shapes of surfaces to be covered vary greatly, perfect fits between discrete units are impracticable, so the joints between adjacent pieces need filling with a moldable material suited to the operating temperatures. Such fill material is designated insulating cement, described in Section IV,D. When good appearance of the outer surface is desired, an additional cement coat containing less coarse ingredients is smoothed over the insulating cement and is known as finishing cement. A major constituent of such cements is asbestos fibers, although many of the insulating cements contain high percentages of mineral wools.

Asbestos fibers are added to mastic materials, such as bitumens, to impart tenacity to the mass, especially after application.

The historical observations of asbestos cloth which resisted fire were carried into the early days of the mandatory "asbestos curtain" in theaters. Here a one-time use was presumed, and the mistaken belief developed that asbestos by itself is an acceptable material for all high fire temperatures. Somehow this error persists. Two factors should be recognized: one, that temperatures of fires in funeral pyres are not especially high, and second, that the duration of the fires was relatively short. As outlined, asbestos does have many uses within the limits of the properties of specific varieties, but prolonged resistance to intense fires is not one of them.

In some installations, loose fibers of asbestos are used to fill cavities or spaces between containing surfaces, but other types of fill insulations usually have property and cost advantages.

Paper containing a substantial percentage of asbestos, so that it is called asbestos paper, is used as wrapping over heated surfaces, or it may be corrugated and formed into tubing for pipe covering, but generally it is not well suited to temperature above about 250F. Neither is it suited to low temperatures because probable wetting will deteriorate it.

The major use of asbestos in the thermal insulation field is as a fibrous ingredient of several types of insulations. It increases physical strength, both useful in handling and for resisting induced forces that cause breakage in service, or for slowing down disruptive breakage during sudden exposures of insulation to temperatures higher than working limits.

2. Reflective Insulations

The first invention of thermal insulation depending upon low emittance and high reflectance is credited to Sir James Dewar. In 1885, he formed a container of thin double glass walls, the Dewar flask. The surfaces of the interposed space were coated with reflective metal, and the space was evacuated to minimize convection.

The Dewar flask is an excellent example of the application of the principle of reducing heat flow between two spaces, the inner bottle and ambient, by minimizing conductance, convection, and radiation. Virtually the only conductance is through the connecting glass at the neck. Convection is reduced by evacuation, and radiation by the low emittance of the outer surface of the inner shell and high reflectance of the inner surface of the outer shell.

Forty years after the invention of the Dewar flask, in 1925, Ernst Schmidt and Edward Dyckerhoff discovered that crumpled tissue-thick aluminum foil sheets, separated except for occasional random contacts, would insulate quite effectively without use of a vacuum. Since that time, commercial insulations depending for their effectiveness upon high reflectance or low emittance have been sold in several countries for use in buildings, principally houses. However, early attempts to use this type of aluminum foil reflective insulation for industrial purposes were not successful.

The first significant reflective insulation suited to power and industrial applications was invented in 1949 by George E. Gronemeyer. Such reflective insulation, without an associated vacuum, may be used for either hot or cold service. At first, the Gronemeyer type of insulation was formed of multiple reflective sheets held separate by nonmetal spacers, which were used because it was presumed that they would provide low conductance from layer to layer. This form was used until about 1955 when it was realized that the low-conductance spacers were limiting the service temperature to a point below that which the reflective sheets could endure. Moreover, the low conductance spacers had to be relatively wide because of their low strength, so that although their heat transfer per unit of area is less than for any metal, the area needed is much greater than for metal spacers. This counteracted the advantage of lower k factor (conductivity) of nonmetal spacers. Since then, the all-metal reflective prefabricated insulation has been adapted to industrial and nuclear power plant needs. Figure 9 shows all-metal pipe insulation. When only aluminum is used, it is suited to about 1000F (540C) service. Higher temperatures can be served with other metals—stainless steel inner layers are suited to temperatures of to 1400F (760C)—but at higher cost.

Fig. 9. All-metal reflective multilayer pipe insulation. Courtesy Mirror Insulation Company.

Other metals can be used for specific properties. The pipe insulation is a simple form. Units are available preformed to fit large and small vessels, elbows, tees, valves, and any shape for which prefabrication is practicable.

The number of layers of reflective metal varies with the service conditions and the efficiency required, each layer adding an appreciable thermal resistance due to surface reflectance of one face and surface emittance of the other. Aluminum is the more practicable metal up to its working temperature, but stainless steel layers are used for higher temperatures until the working temperatures of aluminum are reached. The remaining layers can then be made of the more efficient aluminum. When cost considerations are overridden by specific performance requirements, noble metals may be used.

A specially formed reflective insulation adapted to application between studs of houses has been on the market for several years. In one form, aluminum foils are pleated and covered with other flat sheets so that the internal multiple spaces are bounded by foil. Several such cellular constructions in which the spacing is on the order of one inch or more are sometimes combined to form multiple-layer cellular patterns. For strength, some of the foil sheets are adhered to strong paper for handling; even the more complex forms are shipped with cells folded flat. The assembly is stretched out for application by nailing or stapling the edge strips, reinforced with paper, to the studs or to joists or rafters for under-floor or under-roof constructions. The combinations of foils produce varying thermal resistances (R) for which the producers have published values in their literature.

Gold is an effective reflective insulation with outstanding durability. The spaces between reflective surfaces are usually a substantial fraction of an inch, but a compromise between space dimension and number of layers is found in the so-called superinsulations used in aeronautic space suits, which contain over 100 reflective foil surfaces per inch of thickness. To minimize conductance between the foil layers, very fine fibers of low-conductance material are the separators in some designs.

When practicable, spaces may be evacuated to improve the thermal resistance of the assembly, as in the Dewar flask, but it is important to assure that the evacuation is carried far enough to be worth the effort. The evacuated pressure should be less than about 1/10,000 atm and is often still lower for greater efficiency. At pressures higher than this, so many molecules of air remain in the spaces that through their collisions an appreciable amount of heat is still transferred. In critical cryogenic applications, greater vacuums are used; thermal resistances of 100,000 may be obtained per degree Fahrenheit for one square foot per Btu flow per hour. Literal interpretation indicates that across one inch of thickness a temperature difference of 100,000F is needed to drive one Btu through the material one foot square per hour. More realistically, such insulation an inch thick in a space suit would allow only 3/1000 Btu to pass through one square foot in an hour. If the thickness was only ¼ inch, the heat flow would still be only 1/80 Btu per square foot per hour.

D. Accessories for the Insulation Layer

Many materials are used to install thermal insulations. Those described in this section are associated directly with the insulation materials to improve the insulation layer. Some materials, such as fiberized or filled mastics, are used as insulating coatings.

Cements

Some cements are usable by themselves as thermal insulation, but their efficiency is not as good as many other insulation materials, or they are costly to install because of slow application. Their principal use is as fillers in joints, in repair of small break-outs, or to cover irregular surfaces. Cements that are relatively good insulators are usually coarse grained, so other finer-grained cements are applied over them to improve appearance when desirable, and these are called finishing cements. Although finishing cements provide less thermal resistance than insulating cements, they are expected to be harder and tougher to better withstand operational abuses.

When properties of cements are being evaluated, the effect of rain or unexpected wetting during or shortly after application should not be overlooked, because an appreciable time for setting may be characteristic. Although insulations usually can withstand unexpected wetting, many cements would be washed off. There are locations on which cements are still the more practicable materials, but for standard valves and fittings and standard shapes, preformed fitting covers are preferred. In service all materials designed to be dry must be dry to perform thermally as expected.

In the following list, descriptive densities given are air-dry bulk or apparent values in pounds per cubic foot (pcf).

Material	Temperature Service Range	Inorganic (I) or organic (O)
Alumina-silica fiber. Air setting, dense, 120 pcf.	70 to 2000F (20 to 1100C)	I
Alumina-silica fiber with hydraulic setting binder. Semirefractory. 70 pcf.	70 to 2000F (20 to 1100C)	I
Alumina-silica and asbestos fibers. Air setting, semirefractory.	70 to 1900F (20 to 1040C)	I
Asbestos fibers. Mine run, short fibers. Soft finishing cement. 50–75 pcf.	70 to 1000F (20 to 540C)	I

Material	Temperature Service Range	Inorganic (I) or organic (O)
Asbestos fibers with clay binder. Medium-hard finishing cement or insulating cement 50–70 pcf.	70 to 1000F (20 to 540C)	I
Asbestos, long fibers and binder. Medium-hard high-grade finishing cement and insulating cement. 35–60 pcf.	70 to 1000F (20 to 540C)	I
Calcium silicate with binders. Insulating cement. 20–25 pcf.	70 to 1200F (20 to 650C)	I
Diatomite with binders. Soft, used primarily for high temperatures. 20–30 pcf.	70 to 1900F (20 to 1040C)	I
Magnesia and asbestos fibers. Soft, low mechanical strength. 20 pcf.	70 to 600F (20 to 315C)	I
Mineral wool with clay binder. Medium hard, good thermal resistance, good adhesion to metal surfaces. 20–30 pcf.	70 to 1800F (20 to 980C)	I
Mineral wool with binder and hydraulic set-ting cement. Good thermal resistance, fair adhesion to metals. 30–40 pcf.	70 to 1800F (20 to 980C)	I
Vermiculite with binder. Light, soft, low mechanical strength. Insulating cement. 16–20 pcf.	70 to 1800F (20 to 980C)	I I

E. Materials for Attachment of Installations

Thermal insulations are selected for their suitability to reduce heat transfer when they are applied adjacent to or between specific surfaces. The shapes of these surfaces vary greatly. There may be large flat expanses, tubular sections (capillary to large pipes), curved areas around pipes that are short and large, and around vessels or tanks, irregular intersecting curvatures, or specific mechanical devices as valves, fittings, and transitions that may be so complex and interlocked that some sections are called "Christmas trees." Not only must insulations be attached by means of support suitable for the material while the structure is at normal ambients, but they must also be supported in a mode consistent with the operating motions which temperature changes and service stresses impose, especially when the temperature changes induce moisture migrations. Moreover, the insulation must stop at the structural supports of the constructions being insulated, so that these supports may also need insulations to reduce the "heat leaks," or more importantly, as discussed elsewhere, they may need to be insulated against fire.

The installation in Fig. 1 illustrates a rather simple industrial piping system. Attachment wires over the insulation are pulled tightly so that they become partially embedded in the insulation. Clearance of the insulation on the pipe may be specified in the design, so that diametral expansion of the pipe may take place without putting stresses into the insulation and the attachments. This is accomplished by using a larger inside diameter of the insulation than the outside diameter of the pipe. The sections of insulation may then be pulled tightly against each other; the applicator does not vary the clearance by the amount of tension in the wires for rigid type insulation, as shown. With the softer type of insulations, compression in application may affect clearances if any are needed. Staggered end joints are shown; that is, the two sections have been placed so that the end of one section is about at the midpoint of the other. By this staggering, the applicator handles only one section at a time and wires or bands the bottom of it to the upper portion of the section already placed. Some slight edge roughness which is typical of this type of material will probably be covered with cement before the external jacketing is applied, although it is not necessary from a thermal performance standpoint.

The successful performance of the insulation is partly dependent on the proper attachment and containment of the insulating material. Inadequate supports either reduce the effectiveness of the insulation or lead to premature failure of the insulation system. Moreover, supports of any kind which penetrate the insulation, even though they are slender (as rods or pins) and seem unimportant thermally, may be such large heat leaks that they add as much as 10% to the heat transfer, or affect the envelope adversely. Adverse effects to the envelope include discolorations and embrittlements that develop into cracks at hot spots in high temperature systems, and condensation and corrosion at cold spots in refrigerated systems.

Some materials which are essential adjuncts to thermal insulations are described in the following list. Selection of attaching materials on a simple cost basis without evaluation of the compatibility of the material with a specific insulation often leads to premature and excessive maintenance costs. Generally, the commonly used attachments are not specified in great detail because it is presumed that experienced insulation contractors will use those suitable materials for a particular installation. However, when new materials or unusual requirements for compensation for motion or stress during operations are encountered, specific materials for mounting and the techniques to be used for a particular service must be carefully evaluated, especially when construction codes must be followed.

Many methods of attaching insulations are especially suited to particular materials. Only the more general materials are listed to illustrate the design factors that must be taken into account to produce acceptable insulation systems. Many examples of recent installation techniques, in addition to discussions of thermal insulation properties, are illustrated in two books, one published in late 1969 (7), and the other, less comprehensive, in 1959 (9).

A principal factor of design for attaching insulations is the potential motion of the metal surface (or other material) being insulated. When a steel surface is heated with a 1000F rise, it expands about ¼ inch in 3 feet. Most insulations that are applied for such service shrink slightly at this temperature. The openings thus induced between pieces of insulation can be compensated for by double layers with staggered joints. However, if the insulation is tight on the surface when installed, because of the attachments or simply by friction, the expansion may tear the insulation, at least the first layer. Some preformed insulations are made by cementing two sections of double-layer pipe insulation so that edges are staggered in a small offset, called shiplap.

1. Pipe Insulation Attachments

a. Wires

As shown in Fig. 1, wrapping soft iron wire around the insulation, pulling and tapping to partially embed the wire, and then twisting and embedding the ends is a common method. Wire spacings of about 6 to 9 inches are used.

b. Bands or straps

For pipes of large diameters (usually above 18 inches), light gage steel bands about ⅜ to ½ inch wide, attached with a tensioning tool and clips, provide strong supports for heavy sections of insulation.

c. Pins and Clips

For pipes above about 36 inches, when large curved sections of insulation are to be attached (but only if welding is permitted by codes), special welding pins are attached to the pipe through the insulation, and then slip-on disks or caps hold the insulation. This method supports each piece of insulation individually for single layers.

d. Wrapping and Adhesives

On small pipes, especially for soft fibrous insulation with factory-applied jackets of canvas or reinforced papers, the two halves of the sectional insulation are placed around the pipe and the side and end laps of the jacket adhered with cement furnished with the insulation.

2. Equipment Insulation Attachments

a. Wire Mesh

On large areas, either flat or curved to an arc with a radius of several feet, hexagonal wire mesh (chicken wire) of specified gage of wire, may be placed. In one example, the wire is placed over pins welded to the surface or to special supports on the equipment. Not only does the wire mesh hold the insulation in place but it provides anchorage for plastic cements, which are troweled through and over the mesh.

b. Pins and Clips

Like large pipe insulations, vessel and equipment insulation may be attached by welding pins and clips if permitted by vessel codes.

c. Bands and Straps

On horizontal tanks, in accordance with codes, lugs, bars, or structural shapes are provided so that bands and straps to support the insulation can be installed. These attaching bars or lugs are essential in the lower portion of the vessel to minimize the amount of sagging that develops as the system in service alternates from cold to hot as the equipment is "on and off stream" over the years.

On vertical tanks, vertical support steel, such as angles or channels, is provided at several locations around the circumference so that bands and straps can be installed horizontally as the insulation is applied between the sections of support steel.

3. Pipe Supports

When pipes are supported on structural steel, heat flow from the pipe to the support must often be minimized. Even when insulation can be installed closely around the support, the heat exchange at the many supports may be objectionable for operations. Moreover, a capability for the pipe to slide on its supports may be desirable. For high-temperature services (above 800F), curved blocks of graphite formed to fit specific pipe sizes are available and provide thermal insulation to reduce heat flow below the rate of metal-to-metal contacts. Additional insulation around these blocks further reduces heat transfer rates. For lower temperatures, cradles (preformed supports) under the pipe may be supported on insulation of high compressive strength. Generally, pipes should not be supported directly on insulation because thermal movements and vibration cause adverse abrasion effects leading to premature deterioration.

4. Valve and Fitting Insulation Attachments

Since valves expose a much larger area to heat loss than the length of pipe they replace, complete enclosure of the valve by insulation may be desirable. Consequently, preformed valve and fitting covers (10), rather than application of plastic cements, provide improved system thermal efficiencies and are called for with increasing frequency. Moreover, such preformed covers permit maintenance within the valve without long shut-down of the line. For example, a valve cover usually needs only three bands for attachment, one band at each end and one around the bonnet. For quick maintenance accessibility, removal of the three bands permits separation of the preformed cover into its two halves so that they may be replaced after valve maintenance.

5. Traced Line Attachments

In some pipe systems that transport fluids at elevated temperatures, the fluid must never be allowed to cool below a critical temperature. For example a fluid might solidify at normal temperatures. To minimize the possibility of such serious operating failures, small pipes or tubes are installed within the pipe insulation as shown in Fig. 10. These are heated from independent sources and are called tracers, or heat tracers. They may carry hot fluids or electric heating cables (11). Although only one tracer is shown, two or more may be needed, especially on larger

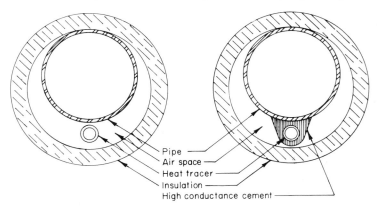

Pipe
Air space
Heat tracer
Insulation
High conductance cement

FIG. 10. Pipe insulation over pipe and heat tracer line, with and without high conductance cement.

pipes or vessels. Since the purpose of the tracers is to transfer heat into the main pipe, the use of a high heat conducting material, as shown in the diagram increases the rate of heat flow appreciably over that which would flow without the special heat-conducting cement. Some designers place the tracer lines above the main lines. This location may have some thermal advantages, but may induce damage if lines are walked on.

6. Cryogenic Temperature Line Attachments

As mentioned above, cold lines must be insulated not only for acceptable heat gain, but even more against moisture ingress. However, note that here "insulated" means either made resistant to heat transfer or made resistant to moisture transfer. Some materials do both. The moisture problem is so complex at cryogenic temperatures that sealed encasements are necessary. Such materials, although associated with thermal insulation, are beyond the scope of this paper. Most of ref. *12* is devoted to cryogenic systems.

After a cold line has been insulated and enclosed with a nonrigid moisture barrier, another layer of insulation with another moisture barrier is sometimes placed over it as an expendable layer, so that damage to it will not affect the performance of the basic insulation system. However, it will improve overall thermal performance even when occasional damage to the outer layer has occurred.

7. Expansion-Contraction Joints

Since the performance of insulations and the surfaces to which or with which they are applied are often contradirectional (the surface expands while the insulation shrinks, even though slightly), systems should include designed means of support and sealing at motion-limiting locations. These are points at which no motion of piping relative to the supports occurs, and may be planned anchorages or inadvertent high friction points. When process systems like that in Fig. 1 are analyzed for motions, it is usually found that the magnitudes within the short tortuous paths are tolerated by the inherent construction joints. However, on long pipes the thermal motions are often so great that special fittings are installed at intervals to enable one section of piping to slip through an enclosing sleeve. Like preformed valve insulation, preformed insulation expansion-contraction fixtures reduce heat flow to the ambient, whereas the induced motion of the insulated surfaces is relatively unrestrained.

F. Envelope Materials

Even when thermal insulation material has been installed, and the joints, cracks, and damaged units have been repaired or cemented, the installations can seldom be considered acceptable at that stage. Even indoors where weather protection is not needed, the appearance would usually be unacceptable to the owner. Moreover, the relatively rough surfaces would be difficult to clean. With few exceptions, thermal insulation materials are covered in some way for one or more of the following reasons:

To keep the insulation free of moisture due to weather.

To keep the insulation free from moisture ingress when surfaces are below ambient dewpoints, which usually indicate the need for vapor-resistant treatments within the construction (see Fig. 11).

To keep the insulation dry when health factors require washing everything within an area with disinfectant.

To keep the insulation dry when process moisture induces dewpoint conditions at different locations within the same area or on different exposed surfaces. Process moisture may be due to natural functions as well as to special processes; for example, moisture may collect on surfaces

Fig. 11. Installing vapor barrier sheet in crawl space under house to avoid condensation in walls and roof. Courtesy Visqueen Div., Ethyl Corporation.

within hot-houses, animal housing of various types, or swimming pool buildings.

To maintain an acceptable appearance for property prestige and owner's pride, or subsequent resale value.

To maintain a pleasing appearance, or to create a color scheme planned for rapid identification of substances within the vessels or pipes, especially useful for operational decisions in emergencies.

To enable dust and airborne grime to be removed readily. Although this cleaning seems to be primarily an enhancement of appearance, it has at least two health-associated aspects. One, dust accumulates to much greater thickness on rough surfaces than it would on smooth, so that when it does fall by its own mass or by accidental disturbances directly or indirectly onto people or animals, it is "dirt." On clothes, it is more aggravating than serious, but in the eyes or lungs, or materials being processed, it can be irreparably damaging. Second, pathological organisms in the dust may accumulate in infectious quantities whereas the normal density may be physiologically rejectable.

To prevent powders or fluids spilled onto the insulation from filtering into fibrous or open pore insulations, or from being sorbed into permeable insulations. In either case, the spillages may create fire hazards which the insulation itself may not possess.

While all additional materials used to cover insulations may contribute some thermal resistances, generally their effects are relatively small, so that their performance thermally is not considered in heat exchange evaluations. However, an important exception is heat reflective aluminum jacketing, as illustrated in Fig. 5 for hot lines, and in Fig. 6 for cold lines.

The three principal classes of coverings over thermal insulation to protect it from weather and from vapor ingress are jackets, coatings, and tapes. Each material must be evaluated for its suitability for a specific service because the exposures are not only to natural ambients, but also to mechanical abuses. Coverings in accessible locations may be replaced easily, so that the economics of frequent replacement of lower cost materials compared with more durable, higher-cost materials should be considered.

An important distinction between weather resistance and vapor resistance should be appreciated. Rain and snow are transient conditions, but flow of vapor may occur continually in accordance with vapor pressure differences (Fig. 4).

The envelope may serve either as a weather shield or as a vapor barrier, or both. The term vapor barrier invites misunderstanding because, just as insulation does not stop heat flow but merely retards the rate of the flow, so a vapor barrier system retards migration of moisture from one side to the other but does not necessarily affect migration of moisture within the barrier envelope. Some vapor barrier materials, especially those in sheet form, may by themselves be impermeable, but the joints and seams and the usual job abuses provide actual or potential leaks. Absolute vapor barriers are most difficult to attain; fortunately, such complete barriers are required only in special constructions. Superior performance of an installed insulation system may result from envelopes which "breathe," but such systems must be designed for local needs.

1. Jackets

Sheet material wrapped around pipe insulation or vessels, either at installation or in a factory, is called a jacket. Frequently used jacket materials are canvas, roll felts, polyethylene, vinyl, and other plastics, and metal such as galvanized steel, aluminum, and stainless steel. The thickness of the materials is specified by the system designer for properties suited to the application needs. Some laminated sheets are made with strong and tough sheet materials combined with others that contribute high resistance to vapor flow, so that the permeance is tolerable.

Roofing and building felts may be suited for weather-resistive jackets, but they are usually of "breathing" permeance rather than good vapor barriers, and therefore selection should be based on permeance tests.

In recent years, when metal jackets are required, preference has been growing for pipe insulation to which aluminum or stainless-steel jackets have been applied at the factory. When selected for a specific climate with consideration of the atmospheric contaminants, they require little maintenance.

Potential hot spots at joints in insulation, caused by high temperatures, are minimized by metal jackets because the high conductance of the metal rapidly transfers heat away from the joints.

2. Coatings

If the jacket itself does not possess sufficient resistance to vapor transmission for a particular service, several plastic materials are available for application by troweling, brushing, or spraying. Some plastic materials are available in colors so that while the coating provides required moisture resistance, it also provides color coding for the materials being transported in the lines, or improves the general appearance of the insulation system. Among plastic materials for coatings are polyvinyl-chloride, polyvinyl-acetate, epoxy, rubber and rubber-like materials, asphalt, and tar. Sometimes addition to the mastics of fibers of asbestos or other minerals, cork, diatomite, or perlite makes the coating suited to direct application to the insulation, but when this is done a mesh reinforcing of glass or metal aids in application and in resistance to induced motions or abuses. Properties of materials must be evaluated, even within a classification, for suitability to withstand weather, chemical fumes and spillage, and fire. Some materials of this general classification are affected by freezing, so that time of year for use must be considered in some localities.

3. *Tapes*

For limited areas and especially around irregular small shapes such as valves and fittings, tapes are used to wrap the insulation. Materials for tapes are rubber and rubberlike plastics, polyethylene, vinyls, and aluminum. Some are reinforced with metal mesh or glass or asbestos fibers. When they are coated with pressure sensitive adhesive, tapes are convenient to apply but may prove to be expensive for large areas.

4. *Sealers and Caulking*

Many openings in insulation systems occur during construction due to accidents and abuse, and also occur in service. Sealing and caulking materials, usually also produced by coatings manufacturers, are adjunct materials that should be kept on hand for normal maintenance.

V. Examples of Premature Performance Problems in Thermal Insulation Systems

Some examples of distresses in thermal insulation systems will be described here to illustrate the need to adhere to the simple principles of heat and moisture flow and to consider them together. The performance of house insulation is probably of widest interest, and several such examples will be cited, but the basic behavior of heat and moisture flow is also illustrated in industrial and commercial constructions. The distresses cited emphasize the simplicity of the principles of thermal insulation. Adhering to them literally, however, may require ingenuity which only experience can give.

A. HOUSE CONSTRUCTION

When a wall in cold climates is insulated, the insulation lowers the outside surface temperature and raises the inside surface temperature, which does seem to be the aim of the insulation. But sometimes the troubles induced may seem to be not worth the cost. Trouble arises due to the lowered outer temperatures, which then are often below the dewpoint, so that moisture *from indoors* condenses within the wall. Moreover, the condensed water frequently freezes. Wet insulation is much less efficient than dry insulation, and in addition, freezing water may generate large compressive forces which can buckle siding on a house if it is flexible, or break it if it is rigid. If the condensed water does not dry before warm weather arrives, wood and other organic materials will decay. Subsurface moisture causes paint to blister and peel.

The accusation: The roof leaked and ruined the carpeting.

In a cold climate, to conserve heat, the walls of the house were made with 2- by 6-inch studs instead of 2- by 4-inch, so that more insulation could be installed. After a severe winter, when the spring thaw arrived, the wall-to-wall carpeting was saturated at the outer walls, presumably due to a leaking roof, although no rain had fallen.

The wet carpet was due to the migration of moisture from indoors through the wall toward outdoors where it condensed and froze. Because of this continuing condensation and freezing, the wall became an ice dam, but since little or no thawing occurred during winter, no water was found on the floor. But during the spring thaw, the ice within the wall melted, ran down the wall and then indoors onto the carpet.

The error: no vapor barrier on or near the interior surface.

The whole roof was a throw-away.

A house had been designed and built properly, and the recommendations of government agencies had been followed. Adequate thermal insulation had been installed between the joists under the roof. One cold windy day, the owner put his head into the crawl space and noticed the cold wind blowing through the space from one louver to another. "This is where my heat is blowing away," he reasoned, obtained boards, and covered the louvers. Two years later he noticed a peculiar odor in the upper rooms, and again peered into the crawl space. The entire space was festooned with fungus growths and all wood surfaces were decaying so badly that the entire roof had to be removed and replaced.

The error: preventing moisture from below, which migrates upward naturally, from escaping to outdoors, for which the properly designed louvers had been installed.

The oversight: the thermal insulation to reduce heat loss was located between the joists *below* the cold air which the owner had noticed blowing *above* the insulation.

The outer walls of the house are badly buckled.

In an insulated wall which contained a good and well-installed vapor barrier, moisture had condensed on the outer board siding and had persisted long enough for the siding to buckle.

The error: the electrician, in installing outlet boxes along the outer wall, cut generous holes into the vapor barrier and then installed *interior* type boxes full of "knockouts." These wide paths for vapor travel into the wall permitted enough moisture condensate to accumulate in the siding to cause buckling.

To avoid the problem: cut vapor barriers carefully, install only *outdoor* type boxes without holes, reset the vapor barrier and tape it to the box without even small openings. Recall the story of the Dutch boy who saved the dike.

Turn around is not fair play.

A house owner decided to finish the interior of a utility room in which a clothes washer and dryer were used, and to install foil-faced insulation in the stud spaces. Despite warnings from neighbors that the foil facing belonged on the inside, he argued that the foil on the outside would reflect summer heat away from indoors. In two years, the odor from the rotting wood and fungus growth was so great that he finally tore out the material and installed new materials properly.

If he had installed the foil-facing in the correct orientation and had left a one-inch space between the foil and the interior wall, the low emittance of the foil would have restrained summer heat from flowing indoors.

What goes up comes down in the wrong places.

A basementless house with crawl space was enlarged and a partial basement added. The excavation uncovered ledge rock carrying so small a quantity of water that the exposed rock was merely damp. When the house was repainted, large blisters formed on a well-insulated wall with vapor barrier. Also, water dripped from the ceiling of an upstairs bedroom which carried insulation above it. The cause: moisture that had been newly exposed evaporated and the vapor flowed horizontally to the colder wall, and up through the stud spaces *behind* the vapor barrier and into the low-clearance roof space. The cold outer siding caused condensation with attendant paint blistering, and the cold roof rafters caused condensation which dripped onto the ceiling insulation and then ran through the ceiling to drip into the bedroom. The remedy: install a vapor barrier on all exposed soil and rock in the under-house crawl space, somewhat as in Fig. 11, and block the wall stud spaces exposed in the crawl space.

B. INDUSTRIAL BUILDINGS

The heat and moisture behavior of commercial buildings is merely a larger scale version of the same behavior in houses.

Heat flow through the roof is excessive.

Heat flow through a large insulated level roof was measured with heat-flow meters and found to be three times as great as expected. The reason for the high heat flow rate was the virtual saturation of the roof insulation. The building was maintained continually at 75F and 40% relative humidity, for which the dewpoint (condensation point) is 47F. Since it was in a northern climate, for an appreciable part of the year the interior of the roofing was below the condensation point. Since the roofing was installed merely as a rain shed, it was open to the air below all along the building walls and also at all penetrations through the roof, as at the many large ducts through the roof. The roof was virtually level but the tiles had sagged into arcs. Condensation on the bottom of the turned-up roofing ran down the curved surface into the insulation. The water ran literally by the laws of hydraulics through the insulation. When the insulation became wet, its efficiency was reduced so that the surfaces of condensation became colder and thereby induced more condensation.

Not enough heat flowed into the roof during warm weather to convert enough liquid water into vapor so that it would dry before cold weather arrived again. Consequently, the insulation remained in different degrees of wetness throughout the year. Three times as much heat flowed through the roof deck system with the insulation wet, as was measured in the adjacent dry roof. Not only did excess heat flow out of the wet roof in winter, but three times as much heat flowed inward during hot weather so that the air-conditioning system was overloaded.

A proposed remedy: seal from below all paths for moisture migration upward which reach to the roofing. Vapor barriers belong on the warm side (inside) and must be tight enough so that any leaks are not of such magnitude that they exceed tolerable limits at points of potential condensation.

Refrigerators are different.

Many refrigerator storage warehouses have troubles that are the reverse of those described above, because the warm side is the outside. As indicated in Fig. 4, vapor pressures at low temperature induce flow from outdoors inward. If the vapor barrier on the warm side is inadequate, moisture will condense and freeze within the insulation, which will reduce its efficiency and therefore induce more ice formation. Since absolute vapor barriers on large structures are seldom achieved, vapor pressure differentials across the insulation layers must be considered. Moisture resistance must not be allowed to arrest vapor flow anywhere, but rather must permit any which gets beyond the vapor barrier to flow completely through the insulated wall and condense on the coils, which can be defrosted. Such action constitutes a design on the "flow-through principle."

VI. Thermal Insulation Systems of the Future

The thermal insulation industry on a substantial scale is only about 50 years old, and many of the important developments have been made in the past 30 years. The foregoing discussion has described the principal thermal insulating materials on the market in substantial volume, the materials associated with them for installation, and the associated materials that constitute the envelope. This should emphasize the need to consider that until the insulation is in place and protected for the service environments, an insulation system has not been constructed. No new principles of physics will be discovered that will affect the performance of insulation systems. Improvements may occur in materials, such as modifications, or the development of entirely new materials that will perform as well or better at lowered life-cost than those now existing. However, new materials in the category of technical breakthroughs seem remote, not that they would not be welcomed, but because synthesized materials seem never to be less costly than existing materials. The new materials of recent years possess higher working temperature capabilities or may require less volume of material for a specific thermal resistance and heat capacity, but some have costs that prevent general adaptability to the mass markets.

If there should be a profound technical breakthrough, the new material would still have to perform in a system of maintainable thermal resistance economically, and be similar in form to the material in use, because thermal insulations are applied materials in the sense that they are adjuncts to other functional constructions.

Improvements in materials and in techniques for application are sought, because installed cost for a required thermal resistance is a major economic factor that dictates the estimated value of the insulation system for an assigned service life.

Plastic materials and composites capable of higher operating temperatures will probably be developed, but shrinkage in long-time service must be less than in most of the currently available materials of this type. Higher working temperatures are mentioned as significant, because when they are exceeded, physical deterioration of the material may make it useless, which is seldom the case for materials selected for lower than ambients and for which thermal efficiency is the principal factor.

Increased handleability of materials is sought because breakage in shipping and handling adds to installed cost.

Resistance to fire is always a factor for which improvement is sought, not only so that the material may remain in place during fire but, even more so, that it may not contribute to the spread or persistance of fire or smoke. Smoke is a hazard not only from toxicity but reduced visibility of persons seeking safe exits or of fire fighters.

VII. Conclusion

This chapter on thermal insulation systems has been a synopsis from which many details have been omitted, but the scope has been wide enough to illustrate that consideration of insulation materials alone is inadequate. An insulation system includes:

1. Insulation materials and the associated cements which constitute the basic heat migration deterrents, but only on a delivered-to-the-job basis. (An insight into the need for "negative" insulations within an insulation system was included in traced lines, Fig. 10.)

2. Other materials, applied not only to hold the insulations in place, but to help them adapt to the stressing motions of the surfaces to which they are applied. These stresses are due to temperature and moisture differentials, vibrations, support restraints, and even abuses due to carelessness.

3. Usually, envelope materials, applied sometimes to protect the insulation, sometimes for health laws, sometimes for appearance, and sometimes for identification of the substance being transported behind the insulated surface, but always to enhance the performance of the insulation system.

Consequently, installed thermal insulation is not complete until it is in a form acceptable to the purchaser as an operable thermal insulation system.

ACKNOWLEDGMENT

Acknowledgment is made for unpublished information on thermal insulations from an industrial consumer viewpoint by W. C. Turner, Staff Engineer, Union Carbide Corporation, South Charleston, West Virginia, as contributed to ASTM

Committee C-16 at various times, especially lists of material classes. Also, some data are from Dr. J. F. Parmer, P. E., as contributed to Vimasco Corporation.

REFERENCES

1. "Book of ASTM Standards," Part 14: Thermal Insulation; Acoustical Materials; Joint Sealants Fire Tests; Building Constructions. American Society for Testing and Materials, Philadelphia, Pennsylvania. (Published yearly, in 33 Parts.)
2. R. P. Tye, "Thermal Conductivity," Vols. I and II. Academic Press, New York, 1969.
3. American Society of Heating, Refrigerating and Air-Conditioning Engineers, Handbooks and Guides: Handbook of Fundamentals, 1967; Systems, 1967; Applications, 1968; Equipment, 1969. ASHRAE, New York. (See comments below.)
4. L. B. McMillan, Heat transfer through insulation in the moderate and high temperature fields. *Trans. ASME* **48**, 1269–1317 (1927).
5. "How to Determine Economic Thickness of Insulation." Nat. Insulation Manufacturers Assoc., New York, 1965.
6. R. L. Boyd and A. W. Johnson, The economics of insulation for electrically heated homes. *AIEE Conf. Paper* **62–401** (1962).
7. J. F. Malloy, "Thermal Insulation," 546 pp. Van Nostrand-Reinhold, New York, 1969.
8. Based on research by Hugh Barty King and Michael Newton for Cape Asbestos Co. Ltd., "Cape Asbestos." Harley, London, 1953.
9. A. C. Wilson, "Industrial Thermal Insulation." McGraw-Hill, New York, 1959.
10. Recommended dimensional standards for prefabrication and field fabrication of thermal insulation fitting covers. ASTM Publication C-450. Am. Soc. Testing and Materials, Philadelphia, Pennsylvania, 1960.
11. R. G. Medley and W. A. Shafer, Electric tracings design simplified. *Hydrocarbon Process. Petrol. Refiner,* **14**(2), 151 (1962).
12. P. E. Glaser *et al.,* Thermal Insulation Systems, A survey. National Aeronautics and Space Administration. NASA SP-5027 (1967).

ADDITIONAL GENERAL REFERENCES

The most useful general references are the current "Handbook of Fundamentals" and other "Guide and Data Books," published every third year by the American Society of Heating, Refrigerating and Air-Conditioning Engineers, 345 East 47th Street, New York, New York, 10017. These Guide and Data Books contain the mathematics of thermal insulation designs and the heating and cooling equipment required to accomplish the environments desired. They also contain the current thermal property data on most materials and combinations of insulations and structural materials in typical constructions. Moreover, the ASHRAE Guide and Data Books contain pertinent literature references. Research by materials producers enables them to introduce improvements in products for the market, and previously published data may be inadequate or obsolete. Consequently, the reader should obtain current product data and recommended practices from producers or their distributors; subsequently, pertinent data appear in ASHRAE Guide and Data Books.

The following references not cited supply some of the earlier background for mathematical analyses of heat transfer. Many are available in most engineering libraries although some may no longer be purchasable. Publications on mathematical procedures for complex transient heat flows are appearing frequently.

13. W. H. McAdams, "Heat Transmission." McGraw-Hill, New York, 1942.
14. M. Jakob and G. A. Hawkins, "Elements of Heat Transfer and Insulation." Wiley, New York, 1942.
15. H. S. Carslaw, "Mathematical Theory of the Conduction of Heat in Solids." Dover, New York, 1945.
16. M. Jakob, "Heat Transfer." Wiley, New York, 1947.
17. H. S. Carslaw and J. C. Jaeger, "Conduction of Heat in Solids," Clarendon Press, London, 1947.
18. L. R. Ingersoll, O. J. Zobel, and A. C. Ingersoll, "Heat Conduction." McGraw-Hill, New York, 1948.
19. G. B. Wilkes, "Heat Insulation." Wiley, New York, 1950.
20. P. D. Close, "Building Insulation." Am. Tech. Soc., Chicago, Illinois, 1951.
21. T. S. Rogers, "Design of Insulated Buildings for Various Climates," An Architectural Record Book, F. W. Dodge Corp., 1951.

AUTHOR INDEX

Numbers in parentheses are reference numbers and indicate that an author's work is referred to although his name is not cited in the text. Numbers in italics show the page on which the complete reference is listed.

A

Akers, L. E., 37
Alexander, J. H., 122(54), *135*
Alton, G., 122(55), *135*
Armstrong, J. R., 183(40), *220*
Ashworth, J. L., 117(82), *136*
Ault, N. N., 103(20), 112(20), *134*

B

Baddour, R. F., 98(13), *133*
Baginski, W. A., 112(31), *134*
Bauer, S. H., 153(13), *219*
Beguin, C. P., 123(62), 125(62), *135*
Bennett, H. E., 238(4), *265*
Beranek, L. L., 58(10), 59(12), 67(10), *86*, 87
Berry, J. M., 176(30), *219*
Betteridge, W., 234(3), *265*
Blocher, J. M., Jr., 107(26), *134*
Blum, S. L., 162(20), 167(20), 168 (20), 173(20), 174(20), 181(20), 182(20), *219*
Bolt, R. H., 54(7), *86*
Bosch, F. M., 126(66), *136*
Boyd, R. L., 294(6), *343*
Bradshaw, W., 183(40), *220*
Brown, A. R. G., 140(6, 7, 8), 142(6), 144(6, 7, 8), 154(6, 7, 8), 162(6), 205(6, 7, 8), *218*
Brumbaugh, J., 37
Buckingham, E., 42(3), *86*
Bunch, M. D., 256(7), 257(7), *265*
Bundy, F. P., 189(52), *221*
Burrows, C. H., 37
Button, D. D., 178(32), 179(32), *220*

C

Cacciotti, J. J., 179(34), 180(34), 205 (61), *220, 221*

Campbell, J. G., 195(56), 198(56), *221*
Carroll, M. N., 37
Carslaw, H. S., *344*
Cauge, T. P., 118(36), *134*
Cawood, L. P., Jr., 256(7), 257(7), *265*
Champetier, R. J., 186(45), *220*
Chanyshev, M. I., 37
Chaussain, M., 238(5), *265*
Chu, T. L., 121(46), *135*
Clark, D., 140(8), 144(8), 154(8), 205 (8), *218*
Clark, T. J., 195(58), *221*
Class, W., 129(73, 74), *136*
Close, P. D., *344*
Coffin, L. F., Jr., 164(23), 165(23), 195 (23), 196(23), 197(23), 198(23), *219*
Courtis, W. F., 92(3), *133*
Cox, G. A., 121(49), *135*
Crede, C. E., 87
Croft, R. C., 213(66), *221*

D

Davern, W., 117(87), *137*
Davidse, P. D., 117(85), *137*
Davis, H. E., 174(28), *219*
Davis, L. W., 104(22), 112(22), *134*
De Groat, G., 130(79), *136*
Den Hartog, J., 87
Deppe, H. J., 37
De Sorbo, W., 184(42), *220*
Diefendorf, R. J., 142(10), 144(10), 157 (10), 164(65), *218, 221*
Donadio, R. N., 168(27), 172(27), 173 (27), 174(27), 175(27), 176(27), *219*
Doo, V. Y., 119(42), *135*
Dorn, J. E., 177(31), *220*
Duff, R. E., 153(13), *219*

345

SUBJECT INDEX

DATE DUE